STEMJUMP

SCIENCE EXPERIMENTS

Written By: Jessica Fan, Carter Feng, Claire Long, Brian Lu, Christopher Ou, Karen Wang, Hannah Xie, Emma Zeng, and Eric Zhang

In Alphabetical Order

Edited By: Brian Lu

ISBN: 9798329444193

Preface

In late 2022, I was one of the captains of the Beckendorff Science Olympiad team, and our team had just come off a very successful first semester. We needed to consider how to fund our second semester of tournaments and travel. I proposed to host some STEM tutoring sessions over winter break to my friends as we are all passionate in STEM. This eventually snowballed into another week of lessons over spring break, and then the founding of STEMJUMP, a non-profit organization, in the summer of 2023.

STEMJUMP's mission is to unleash the potential of young minds by offering the community STEM lessons at an affordable price. In the meantime, giving the high school students the opportunity to build experience by teaching the younger generations.

After the summer sessions of 2023, both virtual and in person, we were left with many experimental ideas and content to show the background knowledge and related applications. So, we created this book to compile the information and experience that our teachers have amassed. I am lucky to have a talented team working with me to create this book and bring to you all our passion for STEM.

I would specifically like to thank Jessica Fan, who initiated the in-person experiment classes in 2023 summer, worked with me to co-found STEMJUMP, and managed many of our activities over the past year.

This book brings fifty-eight easy-to-understand, and fun experiments with related science and engineering background and applications. I hope the readers will enjoy the experiments and information that we have compiled and then marching forward in the journey of exploration in STEM!

STEMJUMP Co-Founder and President, Brian Lu
June 2024

Preface

As a child, I developed a passion for science that has only grown stronger over time. Whether at home or on vacation, I would stare at the ground, searching for rocks, or gaze up at the sky, wondering how rainbows form. During my 7th grade year at Beckendorff Junior High, I was introduced to the Science Olympiad, which was ranked among the top in Texas. Being part of the prestigious team for over a year strengthened my passion and determination in the field of science.

To me, science is more than merely an academic subject in school; it is a beautiful process of inquiring about truth and understanding the world around us. I have since decided to initiate something impactful to share my passion and the beauty of science with others. As a result, my team captain, Brian, and I joined forces to establish STEMJUMP, a non-profit organization. We aimed to inspire the coming generation of STEM students with the beauty and joy of science. As an initial step for our organization, we launched online lectures and science experiment classes in my living room in 2023. Since then, STEMJUMP has continued to engage with students, fostering a passion for science through both online and in-person camps.

This book has been created as part of our ongoing efforts to provide an engaging scientific experience. Our team, including everyone mentioned by Brain, and others, has been invaluable to this process, whether it be creating curriculums, communicating with students, or documenting experiments. I appreciate working with all these dedicated and bright-minded individuals, especially my fellow co-founder, Brian Lu. His commitment to STEMJUMP, coupled with our collective effort, made this book possible.

In this comprehensive guide, our compilation of simple and affordable experiments brings our camps into your home. We encourage you to conduct these experiments with family and friends, creating your own community of scientists and sharing the passion, love and joy.

STEMJUMP Co-Founder and Vice President, Jessica Fan

June 2024

Contents

Supervision by an adult is recommended.

Chapter 1 Biology

1.1 Edible Cell Model by Hannah Xie

Materials:

1. 1 Cupcake
2. 3 Mike and Ikes = mitochondria
3. 4 Skittles Gummies or just normal Skittles = vacuoles
4. 2 Airheads Xtremes = endoplasmic reticulum
5. 4 Gummy worms = Golgi apparatus
6. Sprinkles = ribosomes
7. 1 Oreo = nucleus
8. 4 Sour strips = cell membrane
9. Pack of M&Ms = lysosomes

Procedures:

1. Place one Oreo on the cupcake.
2. Scatter a few Mike and Ikes around the Oreo.
3. Place a few Skittles on the cupcake.
4. Fold Airheads Xtremes into a condensed ribbon form and place it close to the Oreo (make sure they are touching).
5. Get a few gummy worms and place them side by side on the cupcake.
6. Sprinkle some sprinkles on the cupcake.
7. Wrap Sour Strips around the edge of the cupcake frosting.
8. Place a few M&Ms on the cupcake.
9. Observe your cell.
10. Now, you can consume your cupcake. Enjoy!

Safety:

Make sure to wash your hands before preparation and after finishing the experiment. The frosting and candy can get a little messy.

Experiment Questions

1. What are cells?
2. How many cells do you think are in our bodies?
3. Are cells living?
4. Based on previous knowledge, what organelles can you name inside our body's cells?

Background and Applications:

This experiment helps us visualize the building blocks of life: cells. All living organisms consist of one or more cells. Each cell contains a multitude of smaller components called organelles. Each organelle has a specific function that contributes to keeping the cell up and running. Our cupcake represents one animal cell.

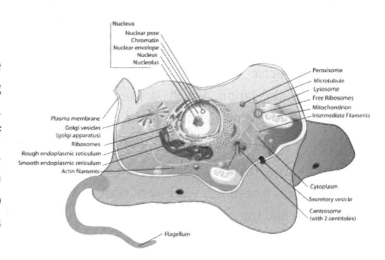

Figure 1.1-1.Diagram of eukaryotic cell. Source: commons.Wikimedia.org

The frosting represents the cytoplasm. The cytoplasm is everything between the cell membrane and the nuclear envelope. This includes things like organelles and cytosol. Cytosol is the jellylike substance that organelles float around in.

Mike and Ikes represent mitochondria, which perform cellular respiration to generate energy. All cells need energy to perform their jobs. Mitochondria take oxygen and glucose as inputs to create ATP, which can be used as energy in an organism. This process also produces carbon dioxide and water as byproducts.

The single Oreo is the nucleus, the brain, and the control center of the cell. The nucleus holds the DNA, genetic information that contains the instructions for cellular functions. DNA in the nucleus is used in special processes called transcription and translation to create proteins in ribosomes. These proteins are then used to carry out certain jobs, like catalyzing another reaction.

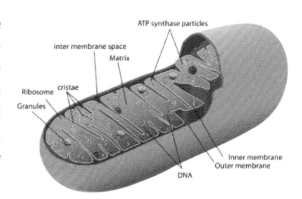

Figure 1.1-2. Diagram of a mitochondria. Source: commons.Wikimedia.org

Skittles represent the vacuoles, which store waste (or water in some organisms). Plants have a large central vacuole that helps to hold up its shape. During periods of drought, vacuoles will contain less water, causing cells to shrink. This is why plants shrivel when dehydrated. When vacuoles are full, they push against the cell wall, causing cells to become turgid (rigid), making plants look upright and healthy. Certain single-celled organisms called protozoa also contain contractile vacuoles. Contractile vacuoles help in osmoregulation, which is the balance of osmotic pressure inside and outside the cell.

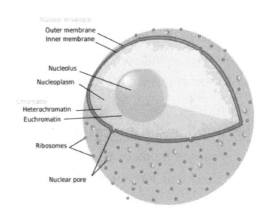

Figure 1.1-3 Diagram of a Nucleus. Source: commons.Wikimedia.org

Sprinkles represent ribosomes. Ribosomes are small organelles that create proteins for usage within the cell. Ribosomes work very closely with DNA inside of the nucleus. Ribosomes utilize transcription and translation to create proteins. Transcription is the process of converting DNA to mRNA (messenger RNA). This mRNA takes the genetic information of DNA outside of the nucleus and towards the ribosomes. Translation is the process of converting mRNA into proteins. mRNA transcribed from the nucleus enters the ribosome. mRNA is composed of codons. Codons are segments of mRNA that contain three adjacent nitrogenous bases. Each codon translates into a specific amino acid. There are 20 total amino acids that combine in different patterns to make proteins. Another type of RNA, tRNA (transfer RNA), gathers amino acids and transfers them (hence the name tRNA) to the ribosome. These amino acids are built up in long polypeptide chains to make a protein.

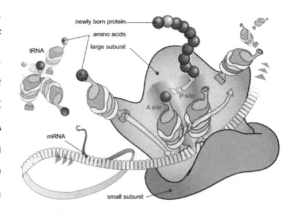

Figure 1.1-4 Complex diagram of a ribosome. Source: commons.Wikimedia.org

Airheads Xtremes represent the endoplasmic reticulum. This organelle can be seen wrapped around the nucleus. This organelle can be classified into three ways: rough, smooth, and transitional. The rough endoplasmic reticulum has many ribosomes, which create proteins for your cells. These proteins are usually used to strengthen/repair the cell membrane. The smooth endoplasmic reticulum has many functions. It can synthesize lipids, metabolize carbohydrates, detoxify the cell, and store calcium ions. The smooth ER also plays a role in the synthesis of sex hormones. The transitional ER creates vesicles (transport bundles) that bud off of the ER. These vesicles contain proteins/other synthesized materials that are carried off to the Golgi apparatus.

The gummy worms represent the Golgi apparatus/Golgi body. The Golgi body carries substances in and out of the cell. They are composed of cisternae (tubular sacs) stacked together. They have a cis and a trans side. The cis side receives vesicles from the ER. Movement of substances through the Golgi apparatus is believed to modify them. In the trans side, vesicles bud off containing the modified substances and are carried away for usage in other areas of the cell or outside of the cell.

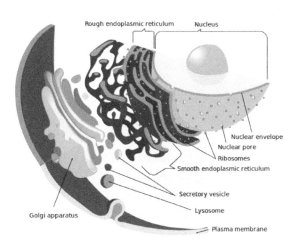

Figure 1.1-5 Diagram of the endomembrane system. Source: commons.Wikimedia.org

M&Ms represent lysosomes, which are mainly present in animal cells. Lysosomes contain digestive enzymes (proteins that speed up reactions) to digest and break down nutrients for the cell.

The Sour Strips wrapped around the cupcake represent the cell membrane/plasma membrane. This membrane keeps all the organelles in the cell and regulates what enters and exits the cell.

Animal cells aren't just restricted to these organelles. They also have centrosomes, which were not represented in today's experiment. Centrosomes are made up of two centrioles. Centrosomes create microtubules that help in cell division. These microtubules will attach to chromosomes in the nucleus during division and help split them up. This process is explained in further detail in the gummy worm mitosis/meiosis experiments.

Today, we explored animal cells, but what about plant cells? We mentioned how plants have large central vacuoles, but did you know they have a cell wall and chloroplasts? Plants contain a tough outer cell wall, not just a flexible cell membrane like animal cells. This cell wall is composed of cellulose and gives plant cells support and protection. Chloroplasts are the maintain photosynthetic organelle of plants. Plants rely on solar energy to create energy and carbohydrates like glucose. Chloroplasts are composed of sacs called thylakoids. These sacs are in stacks called granum. A fluid called stroma fills the space between the thylakoid membranes and chloroplast membranes. Special pigments in thylakoid membranes called chlorophyll absorb light from the sun to promote photosynthesis.

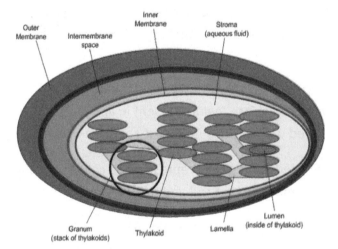

Figure 1.1-6 Diagram of a chloroplast. Source: commons.Wikimedia.org

Cells are extremely intricate even though they are microscopic. A cell's day to day operation is dependent on the harmonious cooperation between many, many organelles. Even to this day, scientists are still discovering fascinating new functions and roles of organelles.

1.2 Fingerprint Visualization by Jessica Fan

Materials:

1. Pencil
2. Paper
3. CLEAR tape

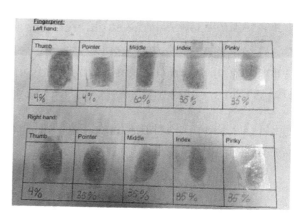

Procedures:

1. Using the pencil, draw a very dark circle on your paper, bigger than your thumb, and color it in as much as you can so that the graphite is powdery on the paper.
2. Put your finger on the circle and press hard.
3. Take your finger off and put tape on it. Be sure you really press the tape on.
4. Now, put the tape on a blank sheet of paper and you should see your fingerprint!
5. Try to identify what type of fingerprint you have.

Safety:

Do not eat the graphite.

Wash your hands before and after the experiment.

Experiment Questions:

1. How many of each fingerprint type do you have?
2. Are there people in the world who don't have fingerprints. If so, why?
3. Which fingerprints do you think would be the rarest?
4. What are some practical uses of fingerprints in the real world?
5. Do you have more similar fingerprints with your family or with your friends? Why do you think that is? Could it be completely random, or are there genetics involved?

Background and Applications:

Have you ever wondered how detectives and cops identify people by their fingerprints? Well, today you get your answer!

You may have heard that every single human being on this planet has a different fingerprint, even twins. Ever wondered why that was? It's not due to genetics, but rather the development of your fingers. When you were a baby, your volar pad (the padding on your fingertips) was covered by embryonic skin composed of layers. These include the epidermis (the outermost layer), the basal layer (situated one layer deeper than the epidermis), and the dermis (located beneath the basal layer). These three layers grow at different speeds. As the dermis develops, the basal layer folds inward, creating bumps that appear on the epidermis as fingerprints. These prints are also determined by pressure changes of the fluids in the womb, meaning that twins won't have the same fingerprints, either.

Now, let's talk about the three basic fingerprint types. (For a more detailed explanation, refer to the section "For Higher Grades"). The three basic fingerprint types are loops, arches, and whorls.

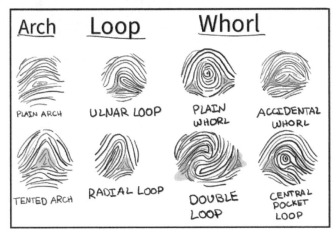

Figure 1.2-1 Fingerprinting types

Loops are the most common type of fingerprint, constituting approximately 65% of all fingerprints. A loop features a single delta, or the small triangles on the fingerprints you see above. The delta is highlighted for your convenience.

Whorls are the second most common type of fingerprint, making up around 30% of all fingerprints. A whorl has two deltas, one on each side of the central circle.

Arches are the rarest fingerprint type, making up about 5% of all fingerprints. There are two types of arches: plain arches and tented arches. Tented arches have a delta at the center of the fingerprint, where the arch rises higher than a normal arch. As you can see in the image above, plain arches do not have deltas.

For Higher Grades:
Moving on to more advanced concepts, we will examine the specific types of fingerprints.

Loops can be differentiated into two types. The ulnar loop opens towards the ulna bone, or the pinky, with the delta positioned toward the thumb. These are the more common types of loops.

The radial loop opens toward the radius bone, or the thumb, with the delta toward the pinky. This type is less common than the ulnar loop.

There are four types of whorls. A plain whorl has a complete circuit/circle in the center of the fingerprint and two deltas on either side. An accidental whorl occurs when the fingerprint combines two different patterns and two or more deltas. A double loop consists of two loop formations with two distinct deltas. Finally, central pocket loops have a small whorl at the center of a loop.

Next, moving on to arches. A plain arch, as mentioned earlier, doesn't have any deltas. The tented arch is characterized by a central delta where the arches rise more prominently compared to a plain arch.

All fingerprint types are special to each person. They are unique identification stamps we carry with us wherever we go. Using these basic lines and ridges on our fingertips, we can unlock devices, perform security procedures, and solve crimes. Who knew something completely up to chance would become such an integral part of our lives?

1.3 Flower Model by Hannah Xie

Materials:

1. Light green construction paper
2. Pink colored paper
3. Dark green construction paper
4. At least 4 Q-tips
5. 1 bead
6. Yellow glitter
7. Tape
8. Scissors
9. Liquid glue

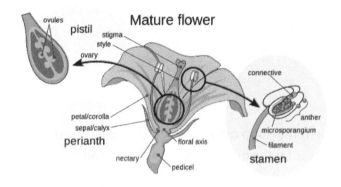

Figure 1.3-1 Diagram of a flower. Source: commons.Wikimedia.org

Procedures:

1. Cut out 6 large petal shaped pieces from your pink paper (long ovoid shapes).
2. Tape them side by side to form a flower (make sure the bottom edge is flat, so it looks like a side profile).
3. Next cut out some smaller petal looking leaves from the light green paper.
4. Tape those to the bottom flat edge of your flower.
5. Cut a long rectangle out from the dark green paper.
6. Tape it vertically in the middle of your flower.
7. Cut out a small circle from the dark green paper.
8. Tape it at the bottom of your long dark green rectangle.
9. Place a small drop of liquid glue onto the center of your circle.
10. Gently place your bead onto the glue and let dry.
11. Now cut one end of the Q-tips so that they are ½ their previous size.
12. Tape 3 on each side of your long green rectangle with the fuzzy end up.
13. Add some liquid glue onto the fuzzy end of each Q-tip.
14. Sprinkle some yellow glitter onto it and let dry.
15. Clean up workspace!

Safety:

Scissors are possibly dangerous so make sure to be careful and ask an adult for help if needed. Make sure to keep liquid glue away from young children. Wash hands carefully after using.

Experiment Questions

1. How many parts of your flower model can you label or identify?
2. Why do you think the petals of flowers are mostly bright colored?
3. What are pollinators? What are some examples?
4. Do you think flowers have genders? Why or why not?

Background and Applications:

The purpose of this experiment is to help us visualize what flowers are like. Flowers play an important role in the reproduction of plants known as angiosperms. Angiosperms are basically flowering or fruit bearing plants. They are one of the more recently evolved clades of plants. Another category is known as gymnosperms. These are cone bearing plants. Plants that came before gymnosperms are known as seedless, vascular plants. Examples include ferns. Lastly, the plants that came before seedless, vascular plants are called bryophytes, or seedless nonvascular plants. These include the liverworts, hornworts, and mosses. In our experiment today, we're primarily going to direct our attention to the reproductive structure of angiosperms, the flower.

Figure 1.3-2 Angiosperm and gymnosperm (respectively). Source: commons.Wikimedia.org

There are many types of plants, so it's sometimes hard to remember what angiosperm means. A quick and easy way to remember is to take the first 4 letters of angiosperm and turn it into the name 'Angie'. Thus, we can say, 'Angie likes flowers'. Flowers are the main reproductive structure in angiosperms. They contain many interesting structures. Let's begin with our petals. Petals are usually brightly colored. Why? This is because it helps to attract pollinators. Pollinators help deliver the pollen to the ovary of the flower. Some examples of pollinators can include bumblebees or butterflies. The small green leaves we cut out and taped to the bottom are known as sepals. These are actually modified leaves. Their hardened texture helps keep the entire flower safe from external damage.

The male reproductive organs in the flower are collectively termed as the stamen. This is what our Q-tip represented. There are two parts to each stamen, the filament and the anther. The anther is the fuzzy part on the top with yellow glitter on top. This is when pollen grains are carried away towards the female reproductive organs. The filament is the stick of our Q-tip. It holds up the anther so it can be more easily accessed by pollinators.

The female reproductive organs in the flower are collectively termed as the carpel. This is the middle section of our plant. Multiple carpels can be grouped into one word, pistil. Like the stamen, our carpel has multiple parts to it. The top of our carpel is called the stigma. This is the top of our long dark green rectangle that we created. The stigma is often sticky, which helps it to trap pollen and start the fertilization process. The next segment is known as the style. This is the long, vertical portion of the carpel below the stigma. The style transports the pollen acquired from the stigma down into the ovary. Lastly, the ovary is the circle that we cut out and taped to the bottom of the long rectangle. The ovary is where ovules, or female sex cells of the plant are stored. Ovules are represented by the bead that we glued into the circle. Pollen from the anther will fertilize the ovules and turn into seeds. This process eventually forms a fruit from the ovary that can disperse the seeds for new growth of plants.

The type of flower often changes amongst species of plants. This is because certain plants must attract different types of pollinators depending on their environment. One way we classify angiosperms is by identifying them as monocots or as dicots. Monocots are angiosperms that have a petal number divisible by 3, their leaf veins are parallel in arrangement, and they have a bundle of shallow, fibrous roots. They are called monocot because they only have 1 cotyledon. Cotyledons are the first leaves that appear when a seed germinates (grows). Dicots are angiosperms that have a petal number divisible by 4 or 5, their leaf veins are branched in arrangement, and they have a deep, central taproot. They are called dicots because they have 2 cotyledons when their seeds germinate.

What we have observed in this experiment is that flowers are composed of many intricate structures that work hand in hand to create beautiful masterpieces of nature that not only look and smell good but also carry out crucial reproductive roles in angiosperms.

1.4 Biology Yeast Balloon by Chris Ou

Materials:

1. Rubber/latex balloons
2. Plastic water bottles (room temp.)
3. Yeast packets
4. Sugar

Procedures:

1. Unscrew the cap of the water bottle.
2. Pour out water until there's about ¼ of the original remaining.
3. Carefully empty the yeast packet into the bottle, make sure not to spill.
4. Pour a teaspoon of sugar into the bottle.
5. Mix gently by swirling or shaking the bottle.
6. Place the neck of the balloon over the opening on the plastic water bottle.
7. Wait 10-20 minutes before observing results.

Safety:

Make sure you dispose the bottles properly, and wash hands accordingly to stay sanitized.

Experiment Questions:

1. What things are yeast often used in?
2. Is yeast a living organism?
3. What do you think would happen if the sugar was left out?

Background and Applications:

The purpose of this experiment is to help us understand how and what yeast exactly is. Once you come back to observe the balloons, you will notice that they have expanded. Seemingly, this air has appeared out of nowhere, however the nature of yeast itself helps explain how and why this reaction occurs.

Yeast, while cheap and generic, serves a vital role in the food industry, most often in baked goods or alcohol. While the process of yeast fermentation in alcohol is interesting, this experiment is more closely related to the rising agent that yeast often is used as in the process of baking bread. In the case of bread, yeast is used because of its unique properties as a living organism. While there are certainly other ways of making bread, using yeast has established itself as one of the most popular methods of baking in the mainstream.

Yeast is taxonomically classified as a fungus, and its scientific name is Saccharomyces cerevisiae, though it is often used as an umbrella term for hundreds of similar species. It grows quickly, and like any other fungi, does well in warm moist environments where it can optimally absorb the most amount of food. They're unicellular but live in multicellular colonies where they reproduce asexually through a process called budding.

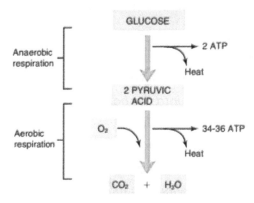

Figure 1.4-1 Aerobic respiration. Source: Earth's Lab, https://www.earthslab.com/physiology/cellular-respiration-importance

$$C_6H_{12}O_6 \ (glucose/sugar) + 6O_2 \ (oxygen) \rightarrow 6CO_2 \ (carbon \ dioxide) + 6H_2O \ (water)$$

In baking, yeast is termed as a, "leavening agent", that is, something that causes dough to rise or expand. The dual processes of aerobic respiration and alcoholic fermentation are especially important in understanding why yeast is able to serve this role so well. While the diagrams below seem relatively complex, we only need to focus on a few key points. "Aerobic", meaning, "in the presence of oxygen", suggests that this process takes place with the presence of oxygen. When there is still oxygen in the bread, yeast uses this process in order to convert glucose (or sugar) into energy. However, the CO2 (Carbon Dioxide) gas byproduct is what leads to the rising of the bread. It allows for the formation of pockets of air that give bread the airy and spongy texture. Without it, bread is dense and doesn't rise.

In the opposing diagram we observe alcoholic fermentation, which occurs when there isn't enough oxygen anymore to support the former reaction. In this process, ethanol is also produced as a byproduct but evaporates during the baking process. The use of both methods means that yeast is a facultative anaerobe, being able to use both processes situationally.

18

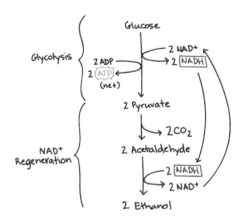

Figure 1.4-2 Alcoholic fermentation/Anaerobic respiration. Source: Khan Academy, https://www.khanacademy.org/science/ap-biology/cellular-energetics/cellular-respiration-ap/a/fermentation-and-anaerobic-respiration

$$C_6H_{12}O_6 \text{ (glucose)} \rightarrow 2C_2H_5OH \text{ (ethanol)} + 2CO_2 \text{ (carbon dioxide)}$$

Now that we have a general understanding of the workings of yeast, let's dive into how it applies to the experiment. We have poured yeast into a moist and ideally warm environment that being the plastic water bottle. We have also given it ample nutrients, like oxygen and glucose, to begin respiration. According to the process above, the yeast uses the available oxygen and begins producing carbon dioxide as a byproduct. Then, it switches to anaerobic respiration, or fermentation, to get the job done, after all the oxygen is used up. This means glucose is directly converted into carbon dioxide and ethanol, and visibly expands the balloon above the bottle with gases.

1.5 Cactus Sponge by Chris Ou

Materials:

1. A light unicolor sponge
2. Scissors
3. A transparent container that can hold water

Procedures:

1. First, get your sponge and cut it into the shape of a cactus.
2. Fill a transparent container with half a cup of water.
3. Place the sponge upright in the container.
4. Check every few hours and observe findings.

Safety:

Make sure to be careful while using scissors. If you are a child, use safety scissors preferably, or have adult supervision!

Experiment Questions:

1. How much of the water is absorbed by the sponge?
2. How are cactuses specifically adapted to living in the desert?
3. What are the similarities between your cactus model and real cactuses?

Background and Applications:

In this experiment, we will be attempting to demonstrate the workings of water absorption in plants. Not only is this vital to fields like botany, dendrology, and biology in general, but is also applicable in real life situations, such as agriscience and farming. Overall, this experiment is intended to give one a better understanding of how plants absorb water and nutrients as a whole.

First, we need some basic knowledge on how plants absorb water. While there are multiple methods, the one we will be focusing on is root absorption, or rather, absorption of food and water by a plant root system. A large misconception is that roots must be in the ground, but the reality is as long as they're attached to something, they can be functional and even grow into other plants/trees. Roots absorb water through a process called osmosis, the natural movement of water from places of high concentration to low concentration, plus the permeability of the roots, allows for the absorption of water. In the case of cacti specifically, aside from having thorns to protect its water, and specialized photosynthesis as well as a hardy outer layer, an important attribute of cacti is its root system. The system is oftentimes very wide, and close to the ground compared to trees or other plants, in order to absorb as much water as possible since desert rains are sporadic and seep through the ground quickly.

The xylem and phloem, present in all vascular plants, are essentially the pathways for which nutrients and water travel. The xylem transports water up from the roots into the plant, while the phloem carries nutrients and other important molecules up and down the plant. These two act like highways in the plant system.

Now let's see how we can apply this information to the experiment. We see the presence of osmosis, as the water slowly is absorbed into the sponge from the water in the container. This closely mimics the osmosis that occurs in everyday root absorption in plants as well. Cacti also in essence act like a sponge. They can efficiently hold and store hundreds of gallons of water, a large amount just like a sponge. In fact, sometimes when you do this experiment, the sponge will suck up all or nearly all of the water in the container, seemingly defying the laws of gravity, just like the osmosis that occurs in plants. As it travels all the way up your cactus stem cutout, we can see that this mimics the pathways, like the xylem, which transports water up from the ground, all the way to the top of the plant.

Figure 1.5-1 Diagram of root systems, left is a cactus, right is a tree. Source: Istock,
https://www.istockphoto.com/vector/big-cactus-growing-next-to-the-tree-gm1343470790-422201040

1.6 Gummy Worm Mitosis by Hannah Xie

Materials:

1. *1 bag of gummy worms (chromosomes)*
2. *7 paper plates (cell)*
3. *1 bag of Licorice Laces (microtubules)*
4. *1 sharpie/marker (nucleus)*
5. *4 Jolly Ranchers (centrosome)*

Procedures:

Prophase: when our genetic information begins to clump

1. *Take 1 paper plate and draw a medium sized circle in the middle (make sure it's big enough so 6 gummy worms can fit).*
2. *Place 6 gummy worms into the circle and arrange them in 3 groups of two (make sure their long sides are touching).*

Prometaphase: when our cell is preparing for division

3. *Take another paper plate and place 6 gummy worms into the middle (arrange them in 3 groups of two, making sure their long sides are touching).*
4. *Take 2 Jolly Ranchers and place them on the right side of the paper plate (make sure they are touching but perpendicular).*
5. *Take 2 more Jolly Ranchers and place them on the opposite side of the paper plate (make sure they are touching but perpendicular).*
6. *Rip the Licorice Lace into small segments and place them in a fan like arrangement in front of both Jolly Rancher pairs.*
7. *Take normal Licorice Lace and place one end at the center of each gummy worm and the other end at the Jolly Rancher pair (gummy worms towards the left should be connected to the Jolly Rancher on the left and vice versa).*

Metaphase: when our cell is aligned and ready to divide

8. *Take another paper plate and arrange the gummy worm pairs so that they're in a vertical line down the middle of the plate.*
9. *Repeat steps 4-7 on the plate.*

Anaphase: when our cell is dividing

    10. *Take another paper plate and repeat steps 4-6.*

    11. *This time, have 3 gummy worms separate of each other on the right and the left. Make sure they are in a vertical line. Use smaller Licorice Lace to connect the centers and the Jolly Ranchers that are closest to it (do this for each gummy worm).*

Telophase: when our cell starts to separate

    12. *Take another paper plate and draw two circles in it.*

    13. *Put 3 gummy worms into each circle.*

Cytokinesis: when our cell completely separates

    14. *Take 2 paper plates and draw one medium sized circle in the center of each.*

    15. *Plate 3 gummy worms into each circle.*

Safety:

Keep small pieces of Licorice Lace and Jolly Ranchers away from small children. This could be a possible choking hazard.

Experiment Questions

1. What do you think the word mitosis means?
2. How long do you think it takes for 1 cell to complete mitosis?
3. Do you think the amount of genetic information or DNA in our cells decreases every time our cells divide? Why or why not?
4. Before cells go through mitosis, their genetic information is replicated. What would happen to the cell's daughter cells if replication did not occur beforehand?

Background and Applications:

The purpose of this experiment is to demonstrate how cells reproduce. Mitosis occurs in the somatic cells of organisms. Somatic cells are any cell that isn't a sex cell (sperm/egg). Mitosis is actually one of the last steps in the process of cell division.

Cell division begins with a long period called interphase. This section is made up of 3 parts. Cell division begins with a period called G1. G1 is when a cell begins to synthesize all of the proteins and substances that it needs for cell division. Organelles will be duplicated and the cell will grow in size. The next stage is the S phase. The S phase is when DNA replication occurs. DNA must have another copy enabled for mitosis to occur, or else each daughter cell created would only have half the number of chromosomes as the parent cell. They would only have half the amount of DNA instructions needed to operate. After the S phase is the G2 phase. This is when more substances are synthesized, and the cell is rearranging its components in preparation for mitosis.

Before we jump into the M phase, or mitosis, we must familiarize ourselves with a few terms. In our experiment, our cell has 6 chromosomes. In reality there are just 3 types of chromosomes, but each is divided to form 6 chromosomes in the S phase. Two chromosomes that are exactly the same are known as homologous chromosomes. Homologous chromosomes in mitosis attach to each other at their centromeres through a protein called cohesin. This is an area in the center of each chromosome that aids in division later on. The chromosomes attached to each other are known as sister chromatids.

Now that we know some basic terminology, let's get into the first phase of mitosis: prophase. During prophase, chromatin condenses into chromosomes. Chromatin is the stringy mass of DNA and proteins that were originally in the nucleus. To properly divide the DNA into 2 daughter cells, the DNA needs to condense into chromosomes. As chromosomes form, the homologs attach to each other at their centromeres to form sister chromatids. This is why we had our gummy worms put into 3 groups of 2. Notice here that we still have a circle surrounding our gummy worms. The circle represents the nucleus, which is where all DNA is stored in a cell.

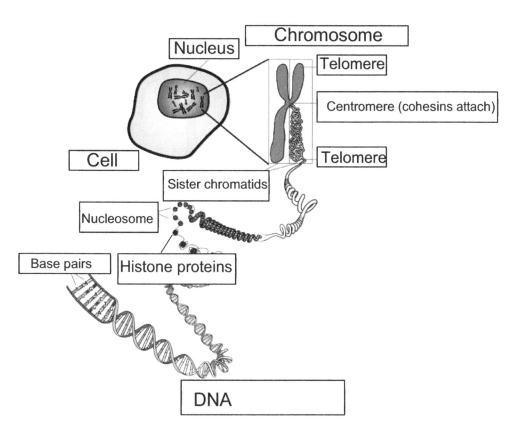

Figure 1.6-1 DNA and its various forms and components. Source: commons.Wikimedia.org

The second phase of mitosis is **prometaphase**. Many things occur here. The circle we originally drew represented the **nuclear envelope/membrane** that normally surrounds our genetic information. The nuclear envelope of the cell fragments during prometaphase, or breaks apart, so you don't need to draw a circle in your paper plate at this stage. During this phase, centrosomes have already positioned themselves at opposite ends of the cell. The Jolly Ranchers are the centrosomes. Each centrosome has 2 **centrioles**. These are organelles in your cell that synthesize **microtubules**, a type of protein filament that is crucial for mitosis. Once they have started to create microtubules, which is what our Licorice Laces represent, an aster is formed. An **aster** is a radial array of short microtubules that fan out from the centrosome. Throughout prometaphase, we begin to see longer microtubules forming. At each chromosomes' centromere is a protein structure called a kinetochore. This is where kinetochore microtubules from the centrosome attach to the chromosome. This is why we had our Licorice Laces connect to the Jolly Ranchers and the gummy worms. During prometaphase, kinetochore microtubules play "tug of war" with each other by pulling back and forth. The sister chromatids do not split because of the cohesin holding them together. Instead, they begin to align. This is important for the next phase: **metaphase**.

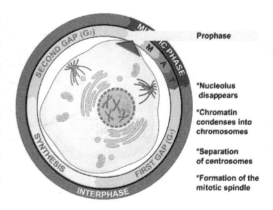

Figure 1.6-2 Prophase. Source: commons.Wikimedia.org

During metaphase, the chromosomes have aligned in a "metaphase plate". This is not a real plate but just refers to the center arrangement of chromosomes. Each sister chromatid is still attached to kinetochore microtubules. **Kinetochores** are areas at the middle of each chromosome that facilitate microtubule binding.

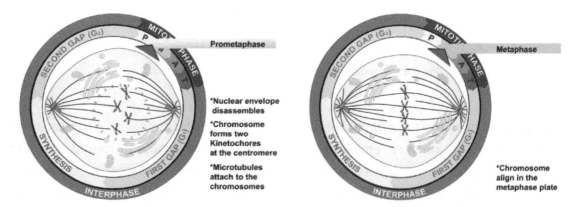

Figure 1.6-3 Prometaphase.   *Figure 1.6-4 Metaphase. Source: commons.Wikimedia.org*

The next phase is anaphase. During this time, cohesin is cleaved/broken by an enzyme called separase. It "separates" the sister chromatids. Each sister chromatid is pulled towards the end that its kinetochore microtubule is attached to. This mechanism is known as the Pac Man mechanism. Special motor proteins will "walk" along the microtubule, dragging the chromatid along with it. The sister chromatid will thus "move". The microtubule that has already been "walked" on by the motor protein will degrade. This gives the impression of something eating off the ends of the microtubule.

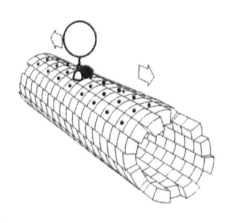

The last phase is known as telophase. This is when the chromosomes decondense. Nuclear envelopes reform around the chromosomes. Chromosomes will loosen, forming **chromatin** (relaxed version of chromosomes). The microtubules formed will all degrade and disappear.

Figure 1.6-5 Motor protein walking along a microtubule, Source: commons.Wikimedia.org

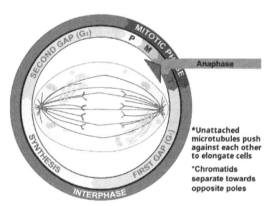

Figure 1.6-6 Anaphase.

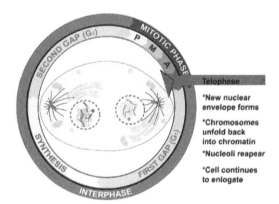

Figure 1.6-7 Telophase. Source: commons.Wikimedia.org

Notice that cytokinesis is not part of mitosis, but it will always accompany it. This is because the cell can't have two nuclei with two sets of DNA. Cytokinesis is the process of splitting the cytoplasm, so the cell turns into 2 separate cells. This is initiated by the formation of a **cleavage furrow**. **Actin microfilaments** (type of protein fiber) will cause the cell to pinch inward, thus splitting the cell into 2 daughter cells. In plants, small vesicles from the Golgi apparatus will align, combine to form a **cell plate**, and then connect with the cell wall, sectioning off the cell into 2 daughter cells.

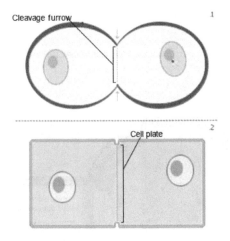

Figure 1.6-8 Cytokinesis in animal and plant cells.   *Figure 1.6-9 Cytokinesis. Source: commons.Wikimedia.org*

Without mitosis, cells would not be able to reproduce. Wounds would not be able to heal, and we would not be able to grow. It's thanks to mitosis that we have many cells and exist at all!

Supervision by an adult is recommended.

1.7 Candy DNA Model by Hannah Xie

Materials:

1. *Pack of gummies*
2. *Toothpicks*
3. *Twizzlers*
4. *Water*

Procedures:

1. *Sort your gummies into 4 colors (blue, red, green, yellow).*
2. *Place 2 Twizzlers parallel to each other with a decent amount of space between them.*
3. *Get 1 toothpick and place 1 blue and red gummy onto it.*
4. *Stick both ends of the toothpick into each side of the Twizzlers.*
5. *Get another toothpick and place 1 green and yellow gummy onto it.*
6. *Stick both ends of the toothpick into each side of the Twizzlers.*
7. *Do this until the toothpicks run across the inside of the Twizzlers until the Twizzlers are full (make sure to only place blue and red gummies together and green and yellow gummies together).*
8. *Hold the top of the Twizzlers arrangement with one hand and the other end with your other hand.*
9. *Twist in opposite directions gently to form a helix shape.*
10. *Clean up workspace!*

Safety:

Toothpicks may be sharp so be sure to keep them out of reach of young children. Don't poke yourself!

Experiment Questions:

1. Have you noticed a pattern in how the gummies are paired?
2. What do you think DNA stands for?
3. What do you think the Twizzlers in our model represent?
4. What do you think the gummies in our model represent?
5. Based on any previous knowledge you have of genes, explain why DNA is so important?

Background and Applications:

The purpose of this experiment is to help us understand what DNA is. You have probably heard and seen DNA before, but why is it so important when it comes to biology? DNA is essentially what determines who we are. It contains the instructions for our cells to create proteins for cellular and bodily function.

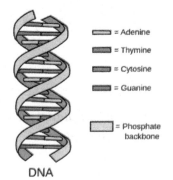

Figure 1.7-1 Diagram of DNA, Source: commons.Wikimedia.org

DNA is made of a few components. This can be easily identified when we take a look at the name of DNA. DNA stands for deoxyribonucleic acid. It sounds long, but in reality, the name is quite literally what it is! DNA is composed of a sugar phosphate backbone. The two sides of the DNA seen in diagrams and represented by our Twizzlers is what connects our nucleotides, which we will talk about later. The phosphates are bonded to the sugar, which is called deoxyribose. There is another sugar, used in RNA, which is called ribose. Deoxyribose is called this because it is lacking an oxygen molecule. The middle of our DNA model is called nucleotides. There are 4 main types of nucleotides, known as guanine, cytosine, adenine, and thymine. There are two main categories of nucleotides, known as the purines and the pyrimidines. Purines are made up of a 5-carbon ring and a 6-carbon ring. Pyrimidines are made up of only 1, 6-carbon ring. The purines are adenine and guanine. The pyrimidines are cytosine and thymine. There is another nucleotide, known as uracil, which we will talk about later. When we constructed our DNA model, you should have noticed that only specific colors could be put next to each other. This reflects something called **Chargaff's rule,** which states that only specific nucleotides can bond together. Adenine can only bond with thymine, guanine can only bind with cytosine. This creates specificity and efficiency in many other cellular processes such as DNA replication and the process of transcription and translation. Each pair of nucleotides is connected through a hydrogen bond.

In the previous paragraph, we mentioned something called RNA. RNA is often depicted as a single strand, rather than a helical double strand, like DNA. One of the biggest differences between RNA and DNA is that one of RNA's nucleotides is different. Instead of having a thymine, it has an uracil. Uracil is still a pyrimidine and will bind to adenine. RNA's sugar in the sugar phosphate backbone is also different. Instead of having a deoxyribose sugar, it has a ribose sugar. Thus, RNA is short for ribonucleic acid. RNA is actually a broad and general term for many types of RNA. There are very specific RNAs that perform a multitude of functions. For example, mRNA forms as an intermediate during transcription and translation which is eventually read and creates proteins. rRNA composes a large percentage of ribosomes in the cells. They also help to synthesize proteins. tRNA also aids in translation to create proteins in a ribosome. Have you seen a pattern? RNAs play an extremely

30

crucial role when it comes to the synthesis(creation) of proteins. Without RNA, many of our cellular functions could not be carried out.

But what about DNA, what roles does it play in organisms? The sole purpose of DNA is to be read into proteins. Proteins catalyze (speed up) reactions in our body and help us function normally. For example, the enzyme lactate is used when we drink milk to digest the sugar lactose. Without it, many people experience mild to extreme gastrointestinal discomfort. DNA makes each of us unique. It determines what we look like, how predisposed we are to certain genetic conditions, and what medical conditions we are born with.

DNA is inherited. It is stored within the nucleus of our cells where it is tightly protected from any outside intrusions by the nuclear envelope. DNA is extremely long. If stretched out, it could reach up to 6 feet in length! Because of this, our cells have found an ingenious way to store this DNA. Usually, when cells are not going through duplication, DNA is stored in a relaxed state called chromatin. This is basically a jumble of DNA and proteins. However, when cells want to divide, DNA is converted into chromosomes. You may have seen these X-shaped complexes before. How does chromatin, a jumble of fibers, become a tightly condensed chromosome? First, DNA is wrapped around histone proteins. These proteins bind and coil DNA around themselves. This continues for a while until our DNA looks like "beads on a string". Then, these histone complexes are further condensed into nucleosomes. Nucleosomes then further condense until an entire chromosome is formed. Condensing DNA into chromosomes makes it much easier for cell division to occur and for the passing of genetic materials to occur.

Each biologically normal human has 46 chromosomes, 23 from their father and 23 from their mother. 2 of these chromosomes are known as the sex chromosomes. You can either have 2 X chromosomes (XX) or 1 X and 1 Y chromosome (XY). XX makes you biologically female and XY makes you biologically male. The rest of the *chromosomes are known as autosomes and code for many other functions in our bodies.*

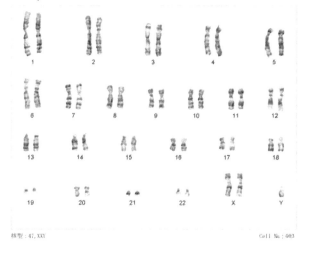

DNA is an important part of us that makes each of us strikingly different but only uses the alternations of 4 specific molecules in a very long strand.

Figure 1.7-2 Human chromosomes, Source: commons.Wikimedia.org

1.8 One Direction Model by Chris Ou

Materials:

1. *Yourself*
2. *A five- or ten-pound weight*

Procedures:

1. *Lift the weight with your dominant arm.*
2. *Perform a standard bicep curl.*
3. *Put your free hand on the upward side of the elbow of the arm performing the curl.*
4. *Repeat and observe carefully.*

Safety:

Make sure you use a weight that you can lift comfortably, and that you do not drop it on your foot or anything potentially fragile.

Experiment Questions

1. How do skeletal muscles operate?
2. What are muscle pairs?
3. What does contracting and relaxing mean?

Background and Applications:

The purpose of this experiment is to help us visualize how exactly skeletal muscles work. While this knowledge is valuable in areas such as bodybuilding and fitness, it is also important to understand how our muscles work in everyday usage with a real-life example. While the bicep curl and the muscle are a generic example that is easy to explain, we will come to learn that every single exercise or action you take has an equal and opposite reaction in an opposing muscle.

Let's first gain a little insight on how our muscles operate. Muscles are subdivided into three distinct categories. Smooth muscles are the ones that move food throughout the body, maintain blood flow, and regulate airways. Cardiac muscle muscles are ones located on the heart, and are used to facilitate the pumping action by contracting the heart. Finally, the most generic type, of which there are hundreds upon hundreds, are the skeletal muscles. This type of muscle will be the focus of this experiment.

Skeletal muscles are attached to bones, allowing the nervous system to facilitate movement and force on an otherwise bare frame of a body. Aside from its primary objective, skeletal muscles also help with stability and posture. The basic setup of a skeletal muscle is one which receives an impulse and contracts, thereby allowing the body to produce movement. However, it is a less well-known fact that muscles always come in pairs, called antagonistic muscle pairs. When one muscle contracts (called the agonist), the opposing will lengthen (called the antagonist)

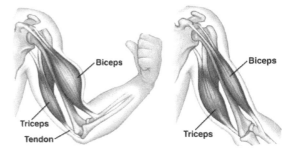

Figure 1.8-1 Agonist-antagonist operation of the biceps and triceps. Source: ResearchGate.net

We will use the diagram above to illustrate this setup. Observing the diagram above, we can easily locate where the bicep (front of the arm), and triceps (back of the arm) are on our body. Tendons, which are often overlooked, serve as the elastic connection between the bone and the muscle, and thus allow contraction to take place. On the left side of the diagram, we see the action of curling your arm upwards will shorten/shrink the bicep muscle back, in technical terms, contracting it. Conversely, as we extend our arm straight out, it is the triceps that shorten/shrink, or contract. The other muscle in each scenario is relaxing, lengthening as the other is shortening. While the biceps and the triceps are the most classic example, countless others such as the hamstrings/quads, pecs/lats, glutes/hip flexors, function in the same manner etc.

Now Putting your hand in between the elbows, you quickly notice a space forming when you curl the weight upwards. This is not due to the bicep muscle shrinking in size, but rather shrinking in length as it contracts. When you straighten your arm back out, you notice that gap is gone, since the bicep has fully elongated and now the triceps, in order to balance out this reaction, is the one that contracts in length.

1.9 Gummy Bear Osmosis by Hannah Xie

Materials:

1. Gummy bears
2. One plastic cup
3. Water
4. Ruler

Procedures:

1. Measure and record the length of your gummy bears from head to toe.
2. Fill the plastic cup with water so that the gummy bears can be completely submerged.
3. Place gummy bears into the water.
4. Wait for 24 hours.
5. Take gummy bears out.
6. Measure and record the length of your gummy bears now from head to toe.
7. Clean up your workspace!

Safety:

It is not recommended that you consume the gummy bears after the experiment is over. Consuming gummy bears before the experiment may be a choking hazard for small children!

Experiment Questions:

1. Was there a difference in the length of the gummy bear before and after soaking in water? Did it increase or decrease?
2. Why do you think this happened to your gummy bear?
3. What did your gummy bear feel like? Were there any physical changes (color, texture, etc.)?

Background and Applications:

The purpose of this experiment is to explain the concept of osmolarity. Osmolarity is the concentration of a specific solute(substance) per liver of solvent (what the solute is in, usually a liquid). Osmolarity is an important fundamental in biology because it regulates the homeostasis(balance) of cells. The survival of cells is important for bodily function.

So, what did we observe in our gummy bears after soaking them in water overnight? If the experiment was conducted correctly, the gummy bears should have increased in size. This is because of osmolarity. When a specific environment has a higher solute concentration, there is less water. When an environment has a lower solute concentration, there is more water. We know that water diffuses(moves) from an area of high to low water concentration, which is why water will always move towards places with higher solute concentration, because there is less water there. In our gummy bears, the concentration of solutes was higher than the surrounding water. The gummy bears contain a very concentrated amount of sugar, which is why it's so sweet. Water contains minerals but is very diluted. This means the water wants to move into the gummy bear to balance out the concentration difference.

The concept of osmolarity, as previously mentioned, is very important when it comes to biology. This is due to the relationship between a cell and its surrounding environment. When the solute concentration is higher outside the cell, we describe this as a hyperosmotic environment. Because solute concentration is higher outside, water from within the cell wants to move out to balance out the external environment. It is diffusing down its concentration gradient. There can be many issues with this. If the external environment is too concentrated, too much water can leave the cell, which could result in plasmolysis. This is when the cell shrinks excessively.

When the cell has a higher concentration of solutes compared to its external environment, this is called hypoosmotic. Water from the cell's external environment will want to diffuse into the cell, down their concentration gradient. This can also be potentially dangerous for many cells. When the cell is too concentrated with solute, and the external environment has too much water, excessive amounts of water can enter the cell. This results in the cell enlarging to a dangerous size. If too much strain is placed on the cell's plasma membrane or cell walls, it could burst. This is called cytolysis.

Lastly, when the cell's internal environment and external environment have equal amounts of solute concentration, there will be no net diffusion of water from either environment.

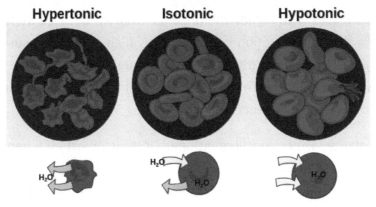

Figure 1.9-1 Osmosis in human red blood cells. Source: Wikimedia.org

Osmolarity plays an important role in all cells in the biological world. We can see examples of osmolarity coming into play in many circumstances. For example, certain types of protozoa (small, eukaryotic, unicellular microbes) have a contractile vacuole. This organelle, which resides inside the cell pumps water out to regulate osmolarity. This can prevent cytolysis. We also see how osmolarity can be applied for the benefit of humans. When we cure meat, we cover it with salt. This creates a hyperosmotic environment around bacteria that could reside in it. This will draw water out of the harmful microbes and kill them, thus disinfecting the meat.

1.10 Heart Model by Hannah Xie

Materials:

1. *3 empty plastic bottles*
2. *1 drill/knife*
3. *4 bendable straws*
4. *Water*
5. *Red food coloring (optional)*
6. *Sharpie*

Procedures:

1. *Drill 1 straw sized hole into a plastic bottle's cap and label the bottle #1.*
2. *Drill 2 straw sized holes in another plastic bottle's cap and label it #2.*
3. *Make sure the cap of the last plastic bottle is off.*
4. *Label the last bottle as #3.*
5. *Arrange the bottles in a row: #1, 2, 3 from left to right.*
6. *Add red food coloring to your water (optional).*
7. *Fill both bottles #1 and #2 with water (#1 should have more).*
8. *Connect 2 straws by sticking one into the bottom end of another.*
9. *Place one end of the straw into the hole in bottle #1.*
10. *Place the other end of the straw into one of bottle #2's holes.*
11. *Connect the last 2 straws by sticking one into the bottom end of the other.*
12. *Place one end of the straw into the remaining hole on bottle #2.*
13. *Place the other end of the straw into bottle #3.*
14. *Squeeze bottle #2 with one hand while pinching the straw connecting #1 and #2.*
15. *Release bottle #2 slowly but instead, pinch the straw connecting #2 and #3.*
16. *Repeat steps 14 and 15!*

Safety:

Make sure an adult is there to drill the holes into the plastic bottle caps. Using sharp objects may result in injuries.

Experiment Questions

1. How many chambers do you think the human heart has? Can you name them?

2. Do you think our heart contracts due to outside nerve input or due to nerve signals inside the heart?

3. The left ventricle is noticeably thicker than all the other chambers. Why do you think this is? (think about where the left ventricle must push blood to)

Background and Applications:

The purpose of this experiment is to help us visualize the transfer of blood in the heart. This model is an oversimplified version of what our heart is actually like. Our heart is composed of 3 layers: the endocardium, myocardium, and epicardium. The myocardium is the thickest layer because it has a large amount of muscle. The heart needs to use this muscle to never stop pumping blood through the vessels in our body. The blood that is pumped contains hemoglobin, a protein that latches onto oxygen, delivering it to our cells. Hemoglobin also carries carbon dioxide away from our cells, back to the lungs to become reoxygenated. The heart contains 4 chambers. The two upper ones are known as the atria (singular atrium) and the ventricles (singular ventricle). To understand this experiment, let's walk through 1 cardiac cycle. Note that everything happening in the pulmonary and systemic circuit occurs at the same time. The atria contract together, and the ventricles contract together. Blood doesn't flow through 1 chamber at a time while the others are empty!

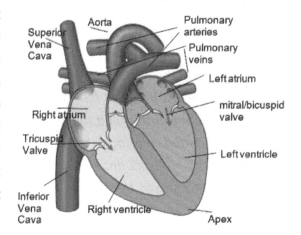

Figure 1.10-1 Diagram of the human heart. Source: Wikimedia.org

Pulmonary Circuit:

Our cardiac cycle begins with the pulmonary circuit, which refers to the lungs. Blood first enters the heart through the superior and inferior vena cava, which are large veins. The superior and inferior vena cava connect to the right atrium. This blood is deoxygenated because it has just come back from visiting the cells, dropping off its oxygen, and picking up carbon dioxide. After the right atrium, the blood flows through valves called the tricuspid valves into the right ventricle. Valves are flaps between chambers and the vessels that connect to chambers, there are 4 in total, that prevent backflow of blood. The tricuspid valve is called tricuspid because it is made up of 3 leaflets. In the right ventricle, the deoxygenated blood is pumped through another valve, called the pulmonary semilunar valve (so called because the leaflets look like half-moons) into the pulmonary trunk. From here, blood flows through pulmonary arteries into the lungs. Blood is then reoxygenated and transported back to the heart through pulmonary veins.

Systemic Circuit:

The systemic circuit is the part of the cardiac cycle that involves transport of blood to the rest of the body. After coming back from the lungs, oxygenated blood is deposited into the left atrium. The left atrium pumps blood through another valve, the mitral or bicuspid valve (so called because it has 2 leaflets) into the left ventricle. The left ventricle is different from all the other chambers in that it has a thicker wall. It has a thicker myocardium. This is because the left ventricle must pump blood to the entire body. This is no easy feat. Blood from the left ventricle is pumped through the last valves, aortic semilunar valves, into the aorta. From here, the aorta branches in many sections into smaller and smaller vessels that deliver oxygenated blood to all the parts of the body. After blood is received by cells, veins pick up carbon dioxide and deliver the blood back to the heart and the cycle begins again.

Now that we have a general understanding of the cardiac cycle, let's dive into our experiment. You might be wondering why we only had 3 plastic bottles. This is because we are mapping out only the systemic circuit. Bottle #1 represents the left atrium. Bottle #2 represents the left ventricle. Bottle #3 represents the rest of the body. When we squeeze bottle #2 and pinch the connection between #1 and #2, we are envisioning the mitral/bicuspid valves closing while the aortic semilunar valve opens. This allows for blood to be pumped into the aorta and towards the rest of the body, which is represented by the filling of bottle #3. When we relax #2 and squeeze the connection between #2 and #3, blood is drawn from bottle #1 into bottle #2. This represents the closing of the aortic semilunar valves and opening of the mitral/bicuspid valves, so blood is pumped from the left atrium into the left ventricle.

Even though this is a simplified model of what our heart is really like, it is hard to imagine the stress that our heart must feel every day. The heart can never stop beating. The coordination between each chamber and each valve allows for harmonious oxygenation and delivery of blood to every cell in the body.

1.11 Spin-A-Round by Chris Ou

Materials:

1. *Yourself*
2. *At least 1 or 2 others*
3. *A chair*

Procedures:

1. *Sit in a chair.*
2. *Stand up from the chair and take 3-5 steps forward.*
3. *Close your eyes.*
4. *Have one or multiple people spin you around a full rotation 10-15 times.*
5. *Open your eyes and slowly navigate yourself back to the chair, but make sure you have people to guide and ensure that you don't fall.*

Safety:

Make sure a trusted person or adult is there to ensure that you safely complete this procedure, since there is a risk of falling! In the chance that you do fall, make sure there are no sharp corners or harmful objects you could hit your head on. Ideally, conduct this experiment on a carpet or soft floor, compared to one that is hard or concrete.

Experiment Questions:

1. How does the human body balance itself?
2. Why do we get dizzy?
3. Why is our vision destabilized even after we've stopped spinning?

Background and Applications:

The purpose of this experiment is to help us understand balance in the human body, a sense that is rarely included among the others. Balance in the human body can be influenced by a variety of factors like sight, touch, etc. However, the system we are going to be focusing on today is the vestibular system. This system is important due to its crucial role in helping our body determine its own position in regards to things around us, helping us, along with our other senses, to navigate the physical world, monitoring position and movement by determining acceleration through a process we will detail below.

Firstly, we must quickly introduce what this system is all about. In humans, the vestibular system is located in the inner ear.

While the exact workings or components of the system are not necessary to understand the experiment, it is important to note the left side of Fig. 1.11-1, which is composed of multiple looping canals. More precisely, there are three of these canals. To put it simply, there is a special fluid inside these canals, which sloshes around, changing how much or which canals they are in, based on the position of the head. If you turn your head or your entire body, the fluid will move around. Small fine-like hair cells are able to sense these changes, and thus convert these fluid movements as electrical signals to the brain that tells the brain how the body is moving.

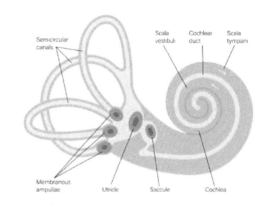

Figure 1.11-1 Diagram of the vestibular apparatus, Source: Premier Neurology

In the experiment, you are asked to spin around many times. You will notice that even after you have stopped spinning, your head and vision still feel blurry, and you feel like you may fall over. This is because your brain is confused. Though you have stopped moving, the fluids in your vestibular system are still moving, meaning that it is sending signals to the brain that you are still spinning, even though you are not. This dissonance is what makes you dizzy or so likely to fall over as you navigate your way to the chair.

1.12 Strawberry DNA by Hannah Xie

Materials:

1. 1 Plastic bag
2. 1/2 Cup of Water
3. 2 plastic cups (1 must be transparent)
4. At least 5 strawberries for best results
5. 2 tsp of Dish soap
6. 1 tsp of Salt
7. Alcohol (ethanol/isopropyl)
8. 1 Coffee Filter
9. 1 Spoon

Procedures:

1. Destem the strawberries and smash them up in the plastic bag (make sure it reaches liquid consistency).
2. Mix 2 tsp dish detergent, 1 tsp salt, and ½ cup water in a separate cup, then pour into plastic bag.
3. Pour the contents in the cup into the plastic bag with the strawberries.
4. Mix around the mixture in bag (make sure not to create too many bubbles from the dish soap).
5. Place a coffee filter over a clean transparent cup and pour out the contents in the bag.
6. Pour an equal amount of alcohol into the filtered strawberry mixture in the cup.
7. Use the spoon to slightly stir the solution.
8. Once a white precipitate starts to form, use the spoon to scoop it up.
9. Make sure to clean up your surroundings.
10. Now you have strawberry DNA!

Safety:

Make sure to keep all your materials in separate containers at the beginning. There will be alcohol and other liquids involved. Keep out of reach of small children and mop up spills immediately.

Experiment Questions

1. What is DNA? What does it stand for?

2. What do you think DNA is made of?
3. Is DNA alive?
4. Try naming the 4 nitrogenous bases of our DNA

Background and Applications:

This experiment helps us visualize and observe DNA, which are the genetic building blocks of all living organisms. If you have cells, you have DNA. So, why did we have to use strawberries for this experiment? Couldn't we have used bananas? Strawberry DNA is relatively easy to extract due to the softness of the fruit (easy to emulsify), the enzymes it contains to help break down its cell wall, and the abundance of genetic material that it harbors. Strawberries are octoploid, which means that they have 8 sets of chromosomes, each containing 7 chromosomes, thus having a total of 56 chromosomes (8x7). This compared to other fruits, like bananas who are also easily crushed but only have 33 chromosomes, makes strawberries relatively easy to work with. Strawberries and other ripe fruits contain enzymes called pectinase and cellulase. Enzymes are specific proteins that help speed up chemical reactions. Pectinase helps break down pectin, which is a polysaccharide in many fruits. Cellulase helps break down cellulose, which largely comprises plant cell walls, thus making it even easier for us to get to the DNA.

Let's start by deepening our understanding of DNA. What is it? DNA stands for deoxyribonucleic acid. Sounds complicated right? In reality, deoxyribonucleic acid tells us the composition of DNA. DNA looks like a double helix. They're composed of something called the sugar phosphate backbone. This is the part you commonly see on the outside of DNA diagrams. The sugar phosphate backbone is made up of phosphate groups attached to the sugar deoxyribose. The inner section of DNA is composed of nitrogenous bases. These are guanine, cytosine, thymine, and adenine. They can be split into two categories, pyrimidines and purines. Pyrimidines are nitrogenous bases that contain a 6-carbon ring. These include cytosine, thymine, and uracil (which we will talk about later). Purines are nitrogenous bases that contain a 6-carbon ring and a 5-carbon ring. These include guanine and adenine. The rule here with DNA is that specific pyrimidines can only pair with specific purines. Thymine can only pair with adenine, and guanine can only pair with cytosine. These nitrogenous bases are attached to the deoxyribose sugars and then form hydrogen bonds with their specific match. This enables us to connect the two sugar phosphate backbones. The sequence of bases in our DNA provides specific and specialized instructions to our cells to perform crucial functions for our survival. Everyone's DNA is different and unique.

Before, we briefly mentioned uracil as also being a pyrimidine. Uracil is also a nitrogenous base, but it only shows up in RNA. RNA stands for ribonucleic acid. This is because the sugar in the sugar phosphate backbone of RNA is not deoxyribose, it's ribose. Uracil replaces thymine and bonds with adenine as a base pair. Typically, you will see RNA as being single stranded instead of double. Now that we have a brief understanding of the genetic building blocks, let's see the mechanisms that helped us to extract them.

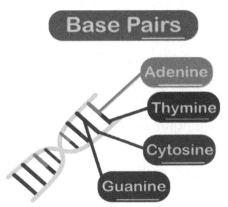

Figure 1.12-1 Basic DNA structure and their nitrogenous base, Source: commons.wikimedia.org

There were 3 main materials that we used in today's experiment: dish soap, salt, and alcohol. Let's begin with the dish soap. Dish soap has the ability to break the phospholipid bilayer of cell membranes and organelle membranes. Phospholipid bilayers are composed of 2 layers of phospholipids (which is a lipid, or a fat, with a phosphate group in its composition). Phospholipids have a hydrophilic (water loving) head and a hydrophobic (water fearing) tail. Thus, the tails remain inside while the heads face the outside, forming a bilayer. Dish soap binds to the hydrophobic ends of the phospholipids, then interacts with water, causing the membrane to disassociate. This eliminates a huge barrier when it comes to extracting DNA. This is because DNA is enclosed in an organelle called the nucleus. It can also be found in other organelles like mitochondria (for more information on organelles, take a look at the edible cell model experiment). Once dish soap disassociates these membranes, DNA can exit the cell, and we can extract it. Next up is salt. Salt eliminates the proteins that bind around nucleic acids. This causes DNA to become loose and free, making it easier for us to collect it. Lastly, alcohol. Alcohol helps the DNA to precipitate, or to solidify. This is the white, gooey, slimy substance that should have appeared at the end of the experiment. DNA is originally hydrophilic, so it can dissolve in water, which is why we can't see it. The alcohol causes it to become more hydrophobic, which is why it precipitates.

What we have explored with this experiment is that DNA is truly fascinating. Without it, our cells would not be able to function and there would be no us!

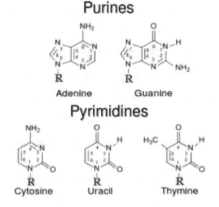

Figure 1.12-2 Chemical structure of nitrogenous bases, Source: commons.wikimedia.org

1.13 Spoons and Grabbers by Hannah Xie

Materials:

1. Multiple plastic spoons
2. Multiple clothespins
3. 2 plastic cups
4. Timer
5. Beads or pom poms
6. LEGO blocks
7. Sharpie

Procedures:

1. Assign 2 teams: one will be the spooners and one will be the grabbers.
2. Label the 2 plastic cups: one with spooners and the other with grabbers.
3. Place the 2 cups at one end of the room.
4. Scatter LEGO blocks all around the room (on the floor, on tables etc.).
5. Tell both teams to line up on the opposite side of the room.
6. Make sure each team member has 1 spoon or 1 clothespin (depending on which team they're on).
7. Start a 5-minute timer.
8. Try to pick up as many LEGO blocks as possible and place them into each teams' respective cups.
9. When the time stops, record how many LEGO blocks each team collected.
10. Clean up the LEGO blocks and replace them with pom poms.
11. Start a 5-minute timer.
12. Try to pick up as many pom poms as possible and place them into each teams' respective cups.
13. When the time stops, record how many pom poms each team collected.
14. Clean up your workspace!

Safety:

If you choose to use beads, keep away from young children. Don't hurt or poke others with your tools (especially the clothespins)!

Experiment Questions:

1. What patterns did you observe from each circumstance?
2. Which team had more LEGOs or pom poms collected?
3. What do you think would happen to the species in this environment if each food source continued to be present? Which species would you think would dominate?
4. How have you seen examples of natural selection in your own life?

Background and Applications:

The purpose of this experiment is to help visualize the process of natural selection. You may have heard of the term "survival of the fittest" and natural selection correlates with this. Natural selection is the phenomenon of animals or organisms with environmental advantages being able to survive and reproduce, thus passing on their genetic information to their offspring. Natural selection plays a key role when it comes to evolution. It allows some species to survive and others to become extinct.

Let's break down today's experiment. We saw two teams: the spooners and the grabbers. We visualize two species of organisms with different food acquisition techniques. First, we began with LEGO blocks as their "food source". The grabbers seemed to have an advantage compared to the spooners at picking up the LEGOs and depositing them into the cup. This means that as time passes, the grabbers will survive because they can eat, and make more offspring. The spooners, who are less capable of spooning up LEGOs, may die out due to starvation.

In the second part of the experiment, however, we switched things up. Instead of having LEGOs as the food source, we changed it to pom poms. This shifted the view of things. Spooners could easily pick up multiple pom poms with each scoop while grabbers were stuck with picking them up one by one. This resulted in the spooners having an advantage against the grabbers. As time goes by, spooners would be able to acquire more resources and survive. They would pass down their genetic information to their offspring. The grabbers, who are disadvantaged, would produce less offspring and possibly face the risk of extinction.

Through this activity, we've seen how one's advantages and disadvantages are purely defined by the environment they live in. Animals have certain adaptations (inherited characteristics) that help them with living in their environment. Adaptations can be applied in multiple scenarios, such as acquisition of nutrients, reproduction, camouflage and so on. Whatever characteristic enables organisms to survive and reproduce can be seen as advantageous.

How have we seen natural selection in the natural world? Let's take a look at the rock pocket mice living in New Mexico. These rodents are an excellent example of natural selection. Their environment contains two types of rock, dark and light colored. Mice living in areas with dark colored rock have a dark coat color. There are few mice living there that have light coat colors. Mice living in areas with light colored rock have light-colored coats. Few mice with dark coats are seen living in those areas. Why does this happen? Because of natural selection. Each coat color is more advantageous for each environment. Mice with coat colors that stand out against their background are more likely to be spotted and preyed on by owls or other predators. Over time, mice with coat colors that matched their background survived more and produced more offspring, resulting in the patterns we see today.

Natural selection occurs over a very long time. This process does not occur overnight. Gradual development and change of populations shift the diversity of organisms in an environment. Because of natural selection, we observe many interesting species that all have special ways in which they adapt to their environment.

1.14 Blind Spot by Chris Ou

Materials:

1. An index card or small piece of paper
2. Marker or pencil

Procedures:

1. Take the index card and
2. fill in a circle on the right side of the long side of the card
3. Draw an X shape on the opposing side
4. Hold the card an arm's length away
5. Close one eye
6. Focus on the cross while slowly bringing the card toward your face

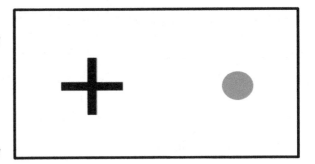

(example card)

Safety:

Make sure you are in an environment with adequate lighting and stop if uncomfortable or if you feel eye strain.

Experiment Questions

1. Why does the circle disappear?
2. What is a blind spot?
3. Does the same effect occur when both eyes are open?

Background and Applications:

Vision is a complicated process, with this experiment demonstrating just one of the optical illusions of many that exist to demonstrate the quirks of vision and how the eye works. The purpose of this experiment is to illustrate the workings of the eye, and why exactly there is a blind spot that we never notice lurking right in the middle of our vision. Eyes are the gateway to our world, and it's quite shocking that smack dab in the middle of that there's a hole that many aren't even aware about. In this experiment we attempt to address these concerns, as well as provide additional information on the specifics.

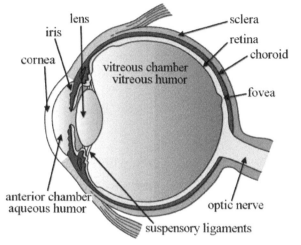

Blind Spot:

A blind spot can be found in all humans, and it's not a health concern. It only is a few degrees wide and tall, and is not particularly interesting itself, but the mechanisms for why it exists are more so.

Figure 1.14-1 Diagram of the human eye. Source: Harvard Eye Associates

The eye is composed of many parts which serve to convert the light that hits our eyes into electrical signals that the brain can process and interpret. The back of the eye is especially interesting. There we have rods and cones. Rods are sensitive to light and are important in night vision and peripheral vision. Cones are important for fine detail, coloration, and central vision. These rods and cones are concentrated in a center in the back of the eyeball called the retina, which ideally is also where the light hitting the eye focuses on if you have perfect vision. Right below is a structure called the optic nerve, which is made up of a pathway of cells that converts what we see into electrical signals, which are then sent up for processing by the brain. These images are then flipped and decoded as the brain projects them for us to see in a format we recognize almost instantaneously.

However miraculous this optic nerve may be, the simple fact that it exists is the cause for the small blind spot in the eye. Since the optic nerve in the back of the eye where it exits the eyeball, it is equipped to convert signals, meaning there are no photoreceptors in that area of the eye. Thus, there's a hole in your vision.

Now that we have a basic understanding of the process that creates a blind spot, let's apply it to our experiment to better understand how it works. When we focus on the cross, and slowly bring the card forward, we are scoping for where our blind spot is. Eventually, we will find it when it matches up perfectly with where the circle is, thereby making it disappear when you focus on the cross because of the optic nerve in the back of your eye.

Finally, we will answer one lingering question you might have. Why have I never noticed it? The answer is simple. We have two eyes. Each of our eyes has an optic nerve, but they create two different blind spots that don't occur in the same location. The brain uses the information from the other eye to fill in the gap for the other. However, this trick doesn't work when we close or cover one eye. This means that it is not noticeable in your everyday vision, and the effect is only possible to achieve when we close one eye.

Chapter 2 Chemistry

2.1 Instant Snow by Karen Wang

Materials:

1. 1 packet of instant snow
2. Water (up to instant snow specifications)
3. Container (optional)

Procedures:

1. Open the instant snow packet and pour it onto the container. The snow will expand a lot so make sure your container is big enough.
2. Slowly pour a little water on the instant snow powder.
3. The snow will instantly start expanding.
4. Play around with the snow. See what happens if you continue adding water.

Safety:

The snow is not edible, so please do not eat it. If any messes are made, clean them up immediately.

Experiment Questions:

1. How do you think the amount of water you add will affect how much snow is produced?
2. Do you think there's a limit to how much water you can add?
3. What causes the snow to be able to expand so much?
4. What's the difference in texture of the snow when you add a little water versus when you add a lot?
5. Where else is instant snow used?

Vocabulary:

Superabsorbent polymer: a polymer that can absorb water up to 300x its weight.

Hydrogen bonding: a type of dipole-dipole force between polar molecules, occurs when Hydrogen is bonded with Oxygen, Nitrogen, or Fluorine, and creates a polar molecule. It is the strongest type of dipole-dipole force, but a lot weaker than covalent and ionic bonds.

Background and Applications:

Superabsorbent polymers:

If you've ever heard of "instant snow", you may think it's some magical substance. As soon as water is added to it, it expands rapidly and forms many crystal looking particles. Wow- magic! But, what's the science behind this?

Instant snow is a substance also known as sodium polyacrylate. Sodium polyacrylate is a polymer (a substance made of long chains of repeating subunits). However, it's a special type of polymer, known as a superabsorbent polymer. When water is added to it, it can soak up 100 to 1000 times its weight!

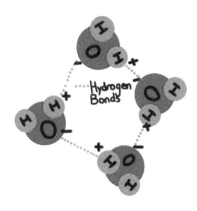

Figure 2.1-1 Hydrogen bond

How is sodium polyacrylate able to absorb so much water? In order to answer this question, we must take a look at what is happening at a molecular level. The molecules of water like to stick to each other using hydrogen bonding. Remember, a water molecule is polar, meaning one end has a slightly positive charge and the other end has a slightly negative charge. The positively charged hydrogen atom of one molecule is attracted to the negatively charged oxygen atom of a different molecule, forming a hydrogen bonding between them. However, in regular water, these bonds aren't very strong.

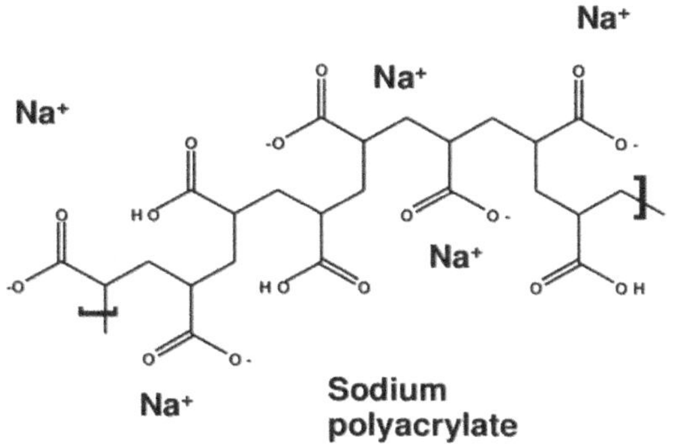

Figure 2.1-2 Structure of sodium polyacrylate. Source: Hubbe, Martin & Koukoulas, Alexander. (2016). Wet-Laid Nonwovens Manufacture – Chemical Approaches Using Synthetic and Cellulosic Fibers. Bioresources. 11. 5500-5552.0.15376/biores.11.2.Hubbe.

Now let's take a look at the structure of sodium polyacrylate. The molecules of sodium polyacrylate are made up of extremely long polyacrylate chains surrounded by sodium ions (Na^+).

When sodium polyacrylate is added to water, the sodium ions like to leave the polyacrylate chains. They want to be equally distributed between the chains and the water. As the sodium molecules leave, water molecules replace their old spots. Water molecules cling very strongly to the side groups in polyacrylate chains, and other water molecules will want to make stronger connections to the ones already connected to the side groups. This causes a lot of water molecules to attach firmly to just one side group. The water molecules will cause the polymer to swell and expand, creating that magical "instant snow".

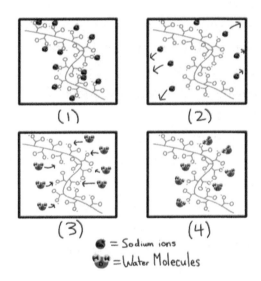

Figure 2.1-3 Interaction between sodium polyacrylate and water.
Image description: (1) A polyacrylate chain initially with sodium ions. (2) When the chains are added to water, the sodium ions like to leave the chains. (3) water molecules then start replacing the places the sodium ions left. (4) Water molecules form hydrogen bonds (yellow) with the negatively charged side groups on the chain.

Superabsorbent polymers have many uses in the real world. For example, they are used in baby diapers. Their ability to hold large amounts of fluids without releasing them makes them an ideal material for many products. They also provide benefits such as keeping the skin dry, protecting against skin irritation, and preventing the spread of infections.

2.2 Skittle Diffusion by Claire Long

Materials:

1. *Skittles*
2. *Round plate*
3. *Hot water*

Procedures:

1. *Form your skittles into a ring/circle.*
2. *Pour the water in the middle of the circle, stopping when it touches the skittles.*
3. *Wait for it to diffuse.*

Safety:

The skittles are still edible, but we advise you not to eat them.

Please do not get burned by the hot water!

Experiment Questions:

1. Why does the color get leached out of the skittles?
2. Why do the colors not mix?
3. What is diffusion?
4. Why does diffusion occur?
5. When and why does diffusion stop?

Background and Applications:

When the skittles are in the water, the coating of the skittles, which is just colored sugar, dissolves. This coating spreads out inside of the water due to a process called **diffusion. Diffusion** is the movement of molecules from areas of high concentration to low concentration. Since areas further inward into the plate have a smaller concentration of molecules, the colored sugar molecules move further and further inwards, until they have reached the center. You notice that the different colored sugar water does not mix: this is because all of these sugar waters have the same exact concentration of sugars inside. They have reached equilibrium: a state of balance. No sugar wants to move to another region of sugar water since there is no area with lower concentrations for it to move into, nor is there any chemical pull. This ensures that all of the colored water stays separate.

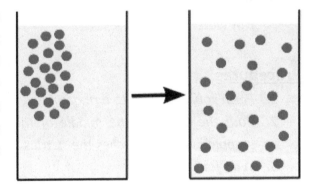

Figure 2.2-1 The diffusion of particles from high to low concentration, eventually reaching equilibrium. Source: https://simple.wikipedia.org/wiki/Diffusion

Diffusion occurs due to a **concentration gradient.** This means that there is a difference in the concentration of particles in one area, and that one area has a high concentration, while another has a lower concentration. Particles move randomly due to their thermal energy, causing them to collide with each other and bounce off of surfaces. These movements are random on their own, but due to many random collisions, the particles begin to change direction and spread out. The increase in particles in higher concentration areas means there are more particles which can be bounced off of, which causes more particles to move into the low concentration area as compared to high concentration area. Eventually, the particles will be spread out evenly, causing a state of equilibrium, and stopping the net movement of particles. Particles still move, but they do not move in a specific direction. This all can be summarized in **Fick's Laws of Diffusion.** The first law states that diffusive flux (an effect that seems to travel through a substance) is related to the gradient of concentration, and that the flux goes from regions of high concentration to low concentration. This is expressed mathematically by $J = -D\frac{d\varphi}{dx}$, in which J is the **diffusion flux**, otherwise the amount of substance that flows through an area during a certain amount of time, measured in amount of substance per unit area per unit time, D is the **diffusion coefficient**, or the rate of diffusion, measured in area per unit time, and $\frac{d\varphi}{dx}$ is the concentration gradient. **Fick's** second law describes how a concentration in a substance changes over time with diffusion, and is mathematically expressed by $\frac{\partial c}{\partial t} = D\frac{\partial^2 c}{\partial x^2}$, in which $\frac{\partial c}{\partial t}$ is the rate of change of concentration over unit time, and $\frac{\partial^2 c}{\partial x^2}$ represents the changes that a concentration can take.

Diffusion is important in many aspects of life. For example, diffusion often occurs in cells. Cells have a semipermeable membrane with a phospholipid bilayer, in which most molecules are not able to pass. Some molecules, given that they are small enough, are able to pass through the bilayer through diffusion. If they can, these molecules are able to dissolve into the membrane, diffuse across, then dissolve in the solution on the other side of the membrane inside of the cell. Note that the phospholipid bilayer is extremely hydrophobic, so only gases, other hydrophobic molecules, and small polar but uncharged molecules can diffuse across the membrane. These molecules follow the concentration gradient, flowing from areas of high concentration (whether the higher concentration area is inside or outside of the cell) to low concentration. During equilibrium, these molecules still move, simply at equal rates in both directions.

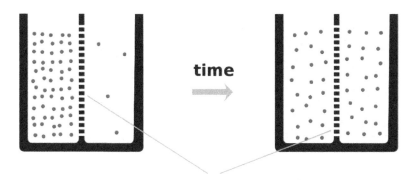

semipermeable membrane

Figure 2.2-2 The diffusion of molecules across a semipermeable membrane with time following the concentration gradient. Source: https://commons.wikimedia.org/wiki/File:Diffusion.en.svg

2.3 Introduction to pH by Karen Wang

Materials:

1. Litmus paper
2. pH chart

Separate cups containing small amounts of the following:

3. Milk
4. Lemon juice
5. Battery acid (optional)
6. Water
7. Baking soda dissolved in water
8. Any other solution you would like to test

Procedures:

1. Dip the litmus paper in the solutions.
2. Observe how the litmus paper changes color.
3. Match the color to the corresponding color on the chart.
4. Repeat for each provided solution to test their pHs.

Safety:

Do not drink battery acid, or any of the solutions. Try not to splash the solutions in your eyes.

Experiment Questions:

Before reading the background information:

1. Why does each solution cause the litmus paper to turn a different color?
2. What specifically causes the litmus paper to change colors?
3. Looking at the solutions that produced red-colored litmus paper, what do you think they have in common?

After reading the background information:

4. How many times greater is the H^+ concentration of a solution with a pH of 2 than a solution with a pH of 4?

Fill out the following table:

Solution	pH	Acidic or Basic?
Baking soda:	9	
Tomato juice	4	
Bleach	13	

Vocabulary:

pH scale: a logarithmic scale ranking the acidity of different substances

Ion: an atom with a positive or negative charge

Basic solution: a solution with a pH above 7

Neutral solution: a solution with a pH of 7

Acidic solution: a solution with a pH below 7

Neutralization reaction: a reaction that occurs when an acid and base react to form water and a salt

Double replacement: a reaction in the form AB + CD -> AD + CB where A, B, C, and D are different elements or compounds

Background and Applications:

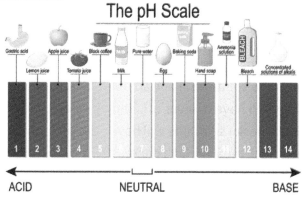

Figure 2.3-1 The pH scale. Source: https://cpo.training/wp-content/uploads/2021/03/860_SS_pH-800x428-1.png

The pH Scale

Everything in life can be classified. We can group objects based on color, shape, size, and many other attributes. We have scales that we use to categorize materials on a spectrum. Although the pH scale may sound like something complicated, it's just one of the few ways scientists classify solutions in chemistry. In simple terms, the pH scale measures how acidic or alkaline (another word for basic, the opposite of acidic) a substance is. You have probably heard of a solution, like lemon juice, being called "acidic" before, but what exactly does this mean?

How do scientists figure out if a solution is very acidic, or not acidic at all?

Scientists use a substance's hydrogen ion concentration when added to water to determine how acidic or basic it is. An **ion** is just an atom with a positive or negative charge. A hydrogen ion has a positive charge, so it is written with the symbol H^+. When solutions are dissolved in water, they will **dissociate** (separate) into positively and negatively charged ions. The more acidic solutions will produce more H^+ ions. But what about the basic solutions? They will produce more hydroxide ions, consisting of 1 oxygen and 1 hydrogen atom and written with the symbol OH^-. Let's look at some examples.

When hydrochloric acid (HCl) is added into water, what do you think will happen? The HCL atoms will dissociate into positive hydrogen ions (H^+) and negative chlorine (Cl^-) atoms. Based on this observation, we can conclude that hydrochloric acid is indeed an acid because it increases the concentration of hydrogen ions.

Let's look at another example, sodium hydroxide (NaOH). When we mix sodium hydroxide into water, it will dissociate into positive sodium ions (Na^+) and negative hydroxide ions (OH^-). We can now tell that sodium hydroxide is a basic solution because it increases the concentration of hydroxide ions. It is very important to not get confused between the terms hydrogen ions and hydroxide ions, even though they sound very similar. Remember, hydroxide ions (OH-) contain hydrogen and oxygen and are negatively charged, while hydrogen ions (H+) are positively charged.

Now that we know what the terms "acidic" and "basic" mean, we have a better understanding of how the pH scale classifies these solutions. Each solution is ranked with a number from 0 through 14. The more acidic solutions will have a lower pH, and the more basic solutions will have a higher pH. Neutral (neither acidic nor basic) solutions will have a pH of 7. An example of this is pure water. The **pH scale** is **logarithmic**, which means that a decrease from one pH level to the next corresponds to a 10x increase in the H^+ concentration. For example, if you had two solutions, one with a pH of 4 and one with a pH of 3, the solution with a pH of 3 would have a H^+ concentration 10x greater than the solution with a pH of 4. The pH of strong acids can be calculated by $pH = -\log[H^+]$. The pH of strong bases can be calculated with the equation $pH = 14 + \log[OH^-]$.

Let's look at the pH of some common substances. Soapy water has a pH around 9, which is basic. Orange juice has a pH around 3, which indicates it is acidic. Distilled water has a pH of 7, which indicates that it is neutral. Try figuring out if the following solutions are acidic or basic based on their pH: (View the table located in the experiment questions)

Now why do scientists study pH so much? Well, that's because maintaining a certain pH level is often very important. Most organisms have a very narrow range of pH their body must stay at in order to survive. For example, our blood has a very specific pH range of 7.35-7.45. An increase or decrease in this level will cause harmful effects on our bodies, such as sleepiness, headaches, loss of appetite, and even death. This is because the enzymes in our blood only operate at very specific pH levels, and a change in these levels would cause them to stop working.

Neutralization Reactions and writing Net Ionic equations

After looking at the basics of pH, we can now dive deeper into the concept of acids and bases, starting with how they react with each other. A **Neutralization Reaction** is a type of **double replacement** reaction. A double replacement reaction is just a reaction where two components of two different molecules essentially "switch places", and they can be written in the form: AB + CD -> AD + CB. The specific definition of a **neutralization reaction** is a reaction that occurs when an **acid** and **base** react to form water and a salt. You can remember it using this form: acid + base -> water + salt.

Now if you're curious, here's how to find the chemical equation and **net ionic equation** of a neutralization reaction (Some background knowledge on chemical equations is required):

Example 1: Write the chemical equation and net ionic equation for the reaction between hydrochloric acid and sodium hydroxide, which is written as: HCl (aq.) + NaOH (aq.)

Step 1. Treat it like a normal double replacement equation. The cations (positively charged ions written in the first part of the compound), hydrogen (H) and sodium (Na) in this case, will switch places. This becomes NaCl and HOH (which can be rewritten as H2O). If you aren't sure what the letters in parentheses mean, aq stands for aqueous, meaning it can be dissolved in water, and I stands for liquid.

$$HCl \text{ (aq.)} + NaOH \text{ (aq.)} \rightarrow NaCl \text{ (aq.)} + H_2O \text{ (l)}$$

Step 2. Balance the equation if needed. This equation is already balanced. Therefore, the above equation represents the chemical equation of this reaction.

Step 3. Now we are going to write the net ionic equation, which ignores the **spectator ions** (an ion that does not participate in the reaction). First, create the complete ionic equation. This is done by separating all strong acids, strong bases, and soluble salts into their respective ions. In our example, HCl is a strong acid, NaOH is a strong base, and NaCl is a soluble salt, so they will all dissociate.

$$H^+ + Cl^- + Na^+ + OH^- \rightarrow Na^+ + Cl^- + H_2O$$

Step 4. Cancel out all ions that are on both sides of the chemical equation. The ions that get canceled out are called **spectator ions**. The equation you are left with is the net ionic equation.

$$H^+ + \cancel{Cl^-} + \cancel{Na^+} + OH^- \rightarrow \cancel{Na^+} + \cancel{Cl^-} + H_2O$$

$$H^+ + OH^- \rightarrow H_2O$$

The ending equation shows us that the pH has become neutralized because there are equal amounts of Hydrogen and Hydroxide ions, and they form water without any extra Hydrogen or Hydroxide ions.

Now try it yourself: Write the full chemical equation and net ionic equation for the reaction between hydrobromic acid (HBr) and potassium hydroxide (KOH).

Your answer should be same as the answer for the example in the text.

2.4 Elephant Toothpaste by Jessica Fan

Materials:

1. ½ cup hydrogen peroxide
2. 1 cup warm water
3. 3 tablespoon yeast
4. Food coloring
5. 1 squirt of dish soap
6. 1 container (plastic bottles work well)
7. 1 plastic funnel
8. 1 aluminum foil tin box

Procedures:

1. Each table will have one set of materials. Place the container inside the tin.
2. Pour the ½ cup of hydrogen peroxide into the container.
3. Add a few drops of food coloring to the hydrogen peroxide. If you want stripes in the toothpaste, put some food coloring around the rim of the container.
4. Add a squirt of dish soap to the hydrogen peroxide and mix. Did anything happen?
5. Mix the yeast and warm water together and wait 1 minute for the yeast to activate.
6. Pour the yeast mixture into the bottle, then step back! This will be hot, so please do not touch it!

Safety:

The elephant toothpaste will be HOT, so please do not touch it.
Wear safety goggles, and be sure to wash your hands before and after the experiment.

Experiment Questions:

1. Why does elephant toothpaste feel hot to the touch?
2. What happens if you add more yeast compared to water, or more water compared to yeast?
3. What happens if you take out the dish soap? How does it change the elephant toothpaste?
4. Why is it necessary to warm up the yeast before performing the reaction?
5. Why do you think this experiment is called "elephant toothpaste"?

Background and Applications:

In science class, you've probably done many cool experiments. But have you ever seen one as cool, or rather, as hot, as this?

In this experiment, you will be able to learn about endothermic and exothermic reactions. First, let's discuss chemical changes. You may already be familiar with the difference between chemical and physical changes, but let's take a quick review. A chemical change occurs when a substance combines with another to create something new. There are five signs of a chemical change, which include: unexpected color change, production of an odor, production of a gas, precipitate formation, and the emission/transfer of energy.

1. Unexpected color change: For instance, mixing two clear liquids and observing the liquid turn purple is an unexpected color change. However, if a purple liquid combined with a clear liquid to produce a light purple liquid, that would not be considered unexpected. Another example would be burning paper, as shown to the right. When paper burns, it starts out white. However, as it burns, it gradually changes color, first to brown, then to black, before finally turning into ashes. You wouldn't expect the paper to change color, but it does!

Figure 2.4-1 Color change in burning paper

2. Production of an odor: Usually, a powder or liquid doesn't produce an odor, or smell. However, mixing them together may result in a smell that wasn't present before. This indicates that a gas was released during the chemical reaction. However, if you mix a liquid with an existing odor with another liquid without an odor and detect a smell, it may not indicate a chemical change. Look at the other signs to be sure!

3. Production of a gas: When two liquids are mixed and begin to bubble, that usually means a chemical change has occurred. Production of a gas can be differentiated from odor, as the odor is something detected by the sense of smell, while the production of a gas is something able to be seen. Production of gas is a tricky sign because boiling a substance also produces gas. Remember, phase changes, such as liquid to gas, are physical changes, whereas unexpected gas production indicates a chemical change.

4. Precipitate formation: A precipitate is a solid created when two liquids react. For example, mixing lemon juice and milk can create curds, which demonstrate a chemical reaction.

5. Emission/transfer of energy: A chemical reaction can produce a change in temperature, emission of light, or both. If mixing two substances results in heat or light, a chemical reaction occurs. It doesn't just have to be heat, however—if a reaction occurs and the temperature of the combined substances drops, that is also an indication of a chemical reaction.

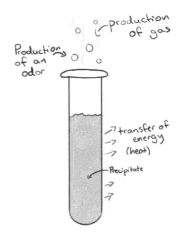

Figure 2.4-2 Gas produced during chemical reaction

Elephant toothpaste grows extremely hot, and it also expands (production of gas), meaning that elephant toothpaste is a chemical change. But why and how does it become so hot? This leads us to the next topic: endothermic and exothermic reactions.

For Higher Grades:

Let's delve into catalysts, endothermic, and exothermic reactions. Catalysts accelerate reactions by lowering the activation energy. For instance, in a reaction that requires 400 units of energy, adding a catalyst could reduce it down to 200. In the case of elephant toothpaste, yeast and warm water act as the catalyst, facilitating the reaction.

There are two types of reactions: exothermic and endothermic. An exothermic reaction releases energy because the products require less energy to create than the reactants. This is demonstrated through the heat release from the elephant toothpaste experiment.

Endothermic reactions absorb energy during the reaction, meaning that the system becomes colder than the surrounding environment. Understanding these reactions helps explain why many chemical reactions may exhibit a rise or fall in temperature.

2.5 Invisible Ink by Karen Wang

Materials:

1. 2 tablespoons of water
2. Two bowls
3. A few drops of Iodine
4. 2 tablespoons of lemon juice
5. 1 cotton swab
6. 1 paintbrush
7. A slip of white paper

Procedures:

1. Pour 2 tablespoons of lemon juice in one bowl.
2. Pour 2 tablespoons of water into the other bowl.
3. Carefully put a few drops of iodine into the water.
4. Dip the cotton swab into the bowl with the lemon juice. Use it to write a message on the paper.
5. Wait for your message to dry. It should disappear once it is dry.
6. After it is dry, dip the paintbrush into the iodine solution and spread a thin layer over the paper. Wait a few minutes.
7. Does your message reappear?

Safety:

Be very careful with the iodine. Wear safety goggles. If any mess is made, clean them up immediately.

Experiment Questions:

1. What is causing the message to reappear?
2. Why does our message stay white while the rest of the paper turns bluish purple?

Vocabulary:

Carbohydrate: 1 of the 4 groups of biological molecules, consists of Carbon, Hydrogen, and Oxygen atoms.

Complementary color: the color produced by the wavelengths of light that are not absorbed.

Charge transfer (CT) complex: an assembly of two or more molecules. Within these molecules, there is an electron acceptor with a partial positive charge, and an electron donor with a partial negative charge.

Electron acceptor: a compound that likes to accept, or take in, electrons from another compound.

Electron donor: a compound that likes to donate, or transfer, its electrons to another compound.

Energy level: refers to the different electron levels in an atom. Each level is located at a fixed distance away from the nucleus, with energy levels increasing as you move further away from the nucleus. Electrons cannot exist between levels, however, but they can move between levels as energy is added or released.

Monosaccharides: the smallest unit of carbohydrates.

Polysaccharides: a long carbohydrate chain made up of many monosaccharides.

Background and Applications:
Starch and Iodine's Color Changing Reaction:

The reaction that causes the paper to turn purple, but leaves your message white after the final step of our experiment, happens because of a chemical reaction between iodine and starch. Starch is a type of **carbohydrate** found in plants. It can also be found in the paper you used to write your secret message. Starch is made up of two types of **polysaccharides: amylose** and **amylopectin**. The polysaccharide responsible for the color changing chemical reaction is amylose. Amylose is made up of many repeating subunits of glucose, a monosaccharide, connected side by side in a long chain. Amylose has a linear shape, with its glucose chain forming a helix (spiral) shape. When iodine comes in contact with amylose, the helix structure enhances the charge transfer between different forms of iodine ions.

Figure 2.5-1 : Iodine atoms (dark circles) inside the amylose helix structure. Source: chemistryviews.org

Now we know how iodine and starch interact, but what causes the paper to turn purple? This color change occurs because of a charge transfer (CT) complex. A **CT complex** is an assembly of two or more molecules. Within these molecules, there is an **electron acceptor** with a partial positive charge, and an **electron donor** with a partial negative charge. Remember, **electrons** are negatively charged particles within an atom. When potassium iodide is added to water, it forms negative iodine ions. These iodine ions can be written as I_3^-, I_5^-, or I_7^-, although I_3^- is most commonly used. The little number on the bottom that differs among the ions tells us how many atoms are in the molecule. The negatively charged iodide acts as an electron donor, and the neutral iodine acts as an electron accepter. During the CT complex, the electrons that are being transferred are constantly moving around because they have lots of energy. As light hits the electrons, they are easily excited to the next highest **energy level**. Once the light hits, it is absorbed by the electrons, and the **complementary color** to this light is emitted. This color happens to be a darkish blue, and it is the same color we observe on the piece of paper.

Where else is this reaction used? This color change between iodine and starch is often used to test substances for the presence of starch in food.

How is our message left white? This is because the lemon juice we used to write the message stops the reaction from occurring. Areas with the lemon juice do not go through the color change.

2.6 Crystal Growing by Claire Long

Materials:

1. Pipe cleaner
2. String
3. Hot water (3 or more cups)
4. Mason jar
5. Popsicle stick
6. Borax: for every cup of water, add 3-4 tablespoons of borax.
7. Food coloring!

Procedures:

Beforehand: Stir a lot of borax into hot water, until it can't dissolve anymore.

1. Bend your pipe cleaner into the shape of your choice.
2. Tie one end of the string to the pipe cleaner and the other to the stick.
3. Place the stick on top of the lid of the jar so that the pipe cleaner is suspended into the mason jar.
4. Carefully Pour the solution of borax and hot water into the mason jar.
5. Optionally, add a few drops of food coloring.
6. Patiently WAIT for the crystals to grow!!! This will take up to 5 days to complete.

Safety:

Please do not drink or eat borax!!!
Eye protection is recommended.
Be careful with hot water!

Experiment Questions:

1. How do crystals form?
2. What is a supersaturated solution?
3. What is a seed crystal?
4. Why is a change in temperature important for crystallization to occur?
5. What would happen if we added an excessive amount of borax, or very little?

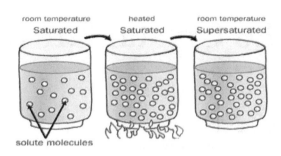

Figure 2.6-1 A supersaturated solution. Source: Collegedunia.com

Background and Applications:

This experiment is an example of a **supersaturated solution**. A supersaturated solution is a solution that contains more than the maximum amount of solute that it can hold at a certain temperature. To explain how this is relevant, consider the properties of borax and hot water.

Borax is a soft crystal that is soluble - it can be dissolved in water. This means that the molecules in borax bond with the water molecules. Additionally, hot water has much faster moving particles, which creates more space and allows for borax to fit within the spaces.

However, when this water cools down, the water molecules get closer together, and all of a sudden, they cannot fit as much borax. Because of this, the borax that was once dissolved now begins to crystallize, often around a **seed crystal**. A **seed crystal** is an object that kick-starts the crystallization process, and is the object that the crystals grow on. In real life applications, a small real crystal is used as a seed crystal, as the solute molecules can easily attach to it, creating a larger, more uniform crystal. In our experiment, this is done by the pipe cleaner: the solvent molecules attach easily to the pipe cleaner, and form a crystalline shape around it.

Seed crystals work since they take advantage of molecular interactions of compounds in a supersaturated solution. The particles in the solution are free to move around randomly, and interact with each other. When the particles inside the solution (borax in our example) collide and interact, they can form a **crystal lattice,** which is the arrangement of molecules that forms a crystalline material. Inserting the seed crystal makes the process much faster, since it removes the need for a slow process that depends on random interactions: instead, the molecules are attached to the seed crystal and interact with and around it, creating a crystal. This is referred to as **nucleation**, which is the transition of solutes from solute to crystal lattice inside of a solution.

2.7 Alka-Seltzer lamp by Karen Wang

Materials:

1. Empty plastic bottle
2. Water
3. Oil
4. Food coloring
5. 1 Alka Seltzer tablet

Procedures:

1. Fill up ⅓ of the bottle with water, and then add your desired choice of food coloring into the bottle.
2. Add twice as much oil as water to your bottle.
3. Break 1 tablet of Alka-Seltzer in half.
4. Add the half of the Alka-Seltzer into your bottle.
5. Watch as colorful blobs start to rise and create your lava lamp.

Safety:

Be careful not to spill the oil or water when pouring it into the bottle. If you do accidentally spill something, make sure to clean it up immediately.

Do not eat any of the materials.

Be careful with splashing the liquids into you or someone else's eye.

Experiment Questions

1. What causes the colorful blobs in the lava lamp to circulate?
2. Is the oil or water reacting with the Alka Seltzer?

Vocabulary:

Density: a measure of how closely together the atoms inside something are packed. It represents the mass per unit volume.

Products: the compounds produced after the chemical reaction.

Reactants: the starting compounds being used in the chemical reaction.

Background and Applications

Introduction to Density:

If you have ever visited a lake or the ocean, you have probably noticed that certain things float on top of the water while other things sink to the bottom. This is because different objects have different densities, or the measure of how tightly packed the atoms in an object are.

Everything around us is made up of very, very, very small units called atoms. A simple way to think of atoms is to picture a bunch of tiny balls floating around in an empty room. Now imagine the balls in one room like to float around very fast, while the ones in the other room like to stay where they are and pack together.

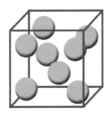 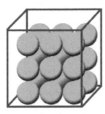

Figure 2.7-1 . The concept of density. The box on the left has a lower density than the box on the right because the molecules are much more spread apart. There is less mass in the first box, therefore it will be less dense. Source: https://theengineeringmindset.com/wp-content/uploads/2015/05/density.png

The faster moving balls are going to be more spread out because they are constantly moving. Therefore, the box on the left will have less mass. On the other hand, the balls in the other room will be much closer together and have a greater mass. Can you imagine if we scooped out a box of balls from each room, which one would float and which one would sink? If you said the box with the tightly packed balls would sink and the other box would float, you're right.

Now we can answer this question: Why does the oil float on top of the water? This is because oil is less dense than water. The previous example gave a basic introduction to what density is. Using our example, we can conclude that if the atoms in a substance are very spread apart, then that substance will float. This means it has a low density. In this case, the atoms in oil are more spread apart than the molecules in water, so the oil is less dense than water. Therefore, it floats on the water.

Exploring Density:

To found out the density of a substance, we can use the formula:

$$Density = Mass/Volume$$

It is also commonly written as p = m/V where p represents density, m represents mass, and V represents volume. The SI (Internation System of Units) unit for density is kilogram per cubic meter (kg/m^3). Density is also commonly measured in grams per cubic centimeter (g/cm^3) for solids and liquids and grams per liter (g/L) for gases. Looking at the formula, we can notice that density is directly proportional to mass. This means that if the mass of an object increases while the volume stays the same, the density will increase too, by the same factor. We can also notice that density and volume have an inverse relationship. This means that if the volume of an object increases while the mass stays the same, then the density will decrease.

Try some examples below of finding density:

1. If I find a rock with a mass of 4.6 grams and a volume of 2.1 cm^3, what is the density of my rock in g/cm^3?

2. Laura wanted to test her luck, so she went to a gold mine to try and dig up some precious gold. At the end of the day, she collected 3 samples that she believed were gold. Sample 1 had a mass of 1.3 grams and a volume of 0.054 cm^3. Sample 2 had a mass of 1.4 grams and a volume of 0.082 cm^3. Sample 3 had a mass of 1.2 grams and a volume of 0.062 cm^3. Which sample was the real gold? (The density of Gold is 19.3 g/cm^3)

3. Belle has a block of wood with a density of 0.45 g/cm^3. She knows the volume of the block is 1.5 cm^3. What is the mass of the block in grams?

Answers: 1. 2.2 g/cm^3 2. Sample 3 3. 1.95 grams

Chemical Reactions:

What causes the bubbles to form in our lava lamp? They form because of a chemical reaction between Alka Seltzer and water. When Alka Seltzer is placed in water, a chemical reaction occurs and carbon dioxide bubbles are produced. You can see these bubbles as they float to the surface of the oil and explode. Now, let's try writing the chemical reaction between Alka Seltzer and water. Every chemical reaction has **reactants** and **products**. The reactants are the starting chemicals being used in the reaction, and the products are the chemicals produced after the reaction. Two compounds can be found in the Alka Seltzer tablets: sodium bicarbonate ($NaHCO_3$) and citric acid. When you dissolve the tablet in water, bicarbonate (HCO_3^-) and hydrogen ions (H^-) are formed. These are the two reactants of the chemical reaction: $HCO_3^- + H^+ \rightarrow H_2O + CO_2$. As the bicarbonate and hydrogen ions collide at the right angle with the perfect amount of energy, they produce carbon dioxide (CO_2) bubbles. These bubbles will then attach themselves to blobs of colored water and float to the surface, creating a lava lamp effect.

2.8 Soap Making by Karen Wang

Materials:

1. Enough soap cubes to fill your mold
2. Food coloring of your choice
3. A soap mold of your choice
4. 1 microwave-safe bowl
5. 1 popsicle stick
6. Optional soap scent

Procedures:

Step 1: Put the soap cubes into a bowl and melt them in the microwave in 30 second intervals until the soap is completely melted. You may need to cut the cubes if they are too large.

Step 2: Add your favorite color and/or scent to the melted soap and use the popsicle stick to mix it together.

Step 3: Carefully pour the liquid soap into the soap mold. The bowl will most likely be very hot, so handle it with caution.

Step 4: Wait for the soap to cool. This will take 24-48 hours.

Step 5: Once the soap has hardened, carefully take it out of the mold.

Safety:

Be sure to have adult supervision when using the microwave. The bowl with the soap will be very hot after it is done melting, so use a cloth for protection when holding the bowl. Do not touch the liquid soap because it's extremely hot. Wear safety goggles during the experiment.

Experiment Questions:

1. What do you think soap is made of?
2. How is soap so effective at cleaning dirt and other things off our hands?

Vocabulary:

Saponification: The process of making soap.

Hydrophobic: molecules that are repelled from water.

Hydrophilic: molecules that are attracted to water.

Polar molecule: a molecule with slightly positive at one end and slightly negative at the other ends.

Non-polar molecule: a molecule without positive/negative ends.

Endothermic reaction: a reaction that absorbs heat from the surroundings.

Background and Applications:

What really is soap?

Soap—it's an essential part of our lives. You most likely use it every day, whether it's for washing your hands or taking a shower. But have you ever wondered how soap actually works? How is it able to take all the dirty germs off your hands, leaving them clean? To answer these questions, we must look closely at the structure of soap. Firstly, what even is soap? Soap is a special type of salt derived from the fats of vegetables, animals, or oil. Some popular soap bases include tallow, coconut oil, and olive oil. This oil/fat base is then combined with an alkaline (having a pH greater than 7) metal solution. Now, let's look at the internal structure of soap.

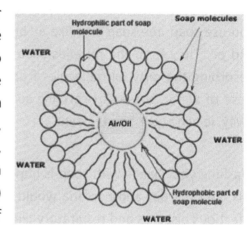

Figure 2.8-1 The internal structure of soap, Source: chemistry.ku.edu.

The image above shows the structure of soap molecules and how they are arranged in a ring shape, so called a **micelle**. As you can see from the diagram above, each soap molecule looks almost like a lollipop, with a head and a long tail. The head of the soap molecule is a hydrophilic polar salt, while the tails are hydrophobic, non-polar fatty acid chains. There were a bunch of new vocabulary words in that last sentence, so let's take a few seconds to understand what they mean. **Hydrophobic** and **hydrophilic** are words describing how attracted to water a molecule is. Hydrophobic contains the prefix "hydro", meaning water, and the root "phobic", which means "fear of something". So you can think of hydrophobic molecules as "water-fearing", as they don't like to go near water. Hydrophilic molecules are the exact opposite. In the word hydrophilic, the root, "philic" means "to love". Likewise, you can think of hydrophilic molecules as "water-loving", as they will be attracted to water. Now, as we see the soap molecules forming a circle, we can understand why the hydrophilic heads are on the outside next to the water, and why the hydrophobic tails are on the inside, away from the water. The next set of words we must understand are "polar" and "non-polar". If a molecule is **polar**, it's basically saying that the molecule has positive and negative charge on two different sides, respectively. To understand this, you can think of the north and south poles of a magnet, one end is negatively charged, and one end is positively charged. As you can see from this diagram of a water molecule (H2O), it is polar. The oxygen is slightly negative, and the hydrogens are slightly positive.

Non-polar molecules are the opposite; they do not have opposite positive and negative ends. Oil

molecules are nonpolar. These nonpolar and polar molecules will repel each other, which is also a reason why oil does not mix with water. Now, we can begin to describe the process of how soap works. When dirty things- like bacteria and dust get on your hands, they mix with the oils on your skin to form dirty oil. When you use soap, the soap acts like a "bridge" between the water and oil. They form the micelles seen in the previous diagram, absorbing the dirty oil molecules from your skin. Then after a rinse of water, these dirt-carrying soap molecules are washed away, leaving your hands clean.

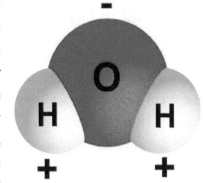

Figure 2.8-2 The polarity of a water molecule. Source: Wikipedia

Washing your hands with soap is important! If you just used water, the oils and water wouldn't mix, and the dirt and bacteria would still stay on your skin. Using soap also reduces the risk of infectious diseases and respiratory illnesses for yourself and others. By keeping your hands clean, you can help prevent the spread of germs to other people.

Saponification: The Reaction Behind It All

The process of making soap is called saponification. More specifically, saponification is the process in which salts of fatty acids (known as soap) and glycerol (a type of alcohol) are formed through the reaction between a base and an ester. An **ester** is a type of chemical compound that is formed when a hydrogen (H) atom in the hydroxyl group (-OH) of an organic acid is replaced by an organic group (-R). The two pictures below show the structure of a carboxylic acid, and the structure of its ester after the hydrogen atom in the -OH group of the carboxylic acid was replaced with an organic group (R').

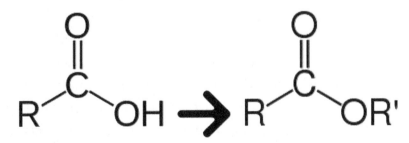

Figure 2.8-3 A carboxylic acid and its corresponding ester. Source: Wikipedia

The saponification reaction can be written in the form: Ester + Base -> Alcohol + Soap.

Let's look at an example of a saponification reaction between triglyceride and sodium hydroxide (NaOH). Triglyceride is the ester and sodium hydroxide is the base. This is the reaction that occurs:

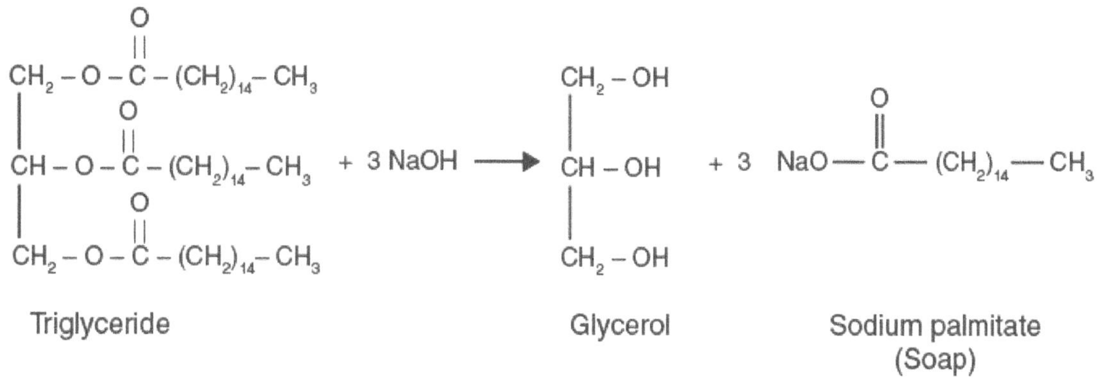

$$CH_2-O-\overset{\overset{O}{\|}}{C}-(CH_2)_{14}-CH_3$$
$$CH-O-\overset{\overset{O}{\|}}{C}-(CH_2)_{14}-CH_3 \quad +\ 3\ NaOH \longrightarrow$$
$$CH_2-O-\overset{\overset{O}{\|}}{C}-(CH_2)_{14}-CH_3$$

$$CH_2-OH$$
$$CH-OH \quad +\ 3 \quad NaO-\overset{\overset{O}{\|}}{C}-(CH_2)_{14}-CH_3$$
$$CH_2-OH$$

Triglyceride   Glycerol   Sodium palmitate
                                                          (Soap)

Figure 2.8-4 The reaction between an ester, triglyceride, and sodium hydroxide. Source: https://byjus.com/chemistry/saponification/

Another commonly used base in making soap is potassium hydroxide (KOH). Soaps derived from sodium bases produce "hard" soaps while soaps derived from potassium bases are "soft" soaps.

The **saponification value** of a base refers to the amount of the base that is needed to saponify the esters. It is commonly listed in milligrams of KOH.

Saponification is not only used in making soaps, but also used by fire extinguishers. To convert fats and oils into non-combustible soaps that can be used to extinguish fires, wet chemical fire extinguishers use the saponification process. Another good thing about this reaction is that it is **endothermic** (the reaction absorbs heat from the surroundings) and can lower the temperature of the flames.

2.9 Incredible Ice Cream by Karen Wang

Materials:

1. 1 tablespoon of sugar
2. ½ cup of half and half
3. ¼ teaspoon of vanilla extract
4. ⅓ cups of rock salt
5. 1 Pint-sized storage bag
6. 1 Gallon-sized storage bag   3 cups of Ice cubes or crushed ice
7. A towel to prevent your hands from getting too cold

Procedures:

1. Pour the half-and-half, vanilla extract, and sugar into the smaller bag.
2. Gently shake the bag around to mix the ingredients. Make sure the bag is closed tight!
3. Add the ice cubes and salt to the larger bag.
4. Place the smaller bag into the larger bag. Wrap the bags with a towel to prevent your hands from getting too cold. Make sure both bags are closed very tight, and start shaking! (Do not shake too aggressively).
5. Stop when the mixture in the smaller bag begins to solidify. (This will take about 10-15 minutes)
6. Enjoy your delicious treat!

Safety:

Do not shake the bag too hard! Otherwise, the smaller bag will pop, and you will have salty ice cream. If anything is accidentally spilled, immediately clean it up. Make sure to clean up your area after you are done making the ice cream.

Experiment Questions

1. Why did we need to add salt to the ice?
2. What role does salt play?

Vocabulary:

Ion: An atom that has either gained or lost an electron and no longer has a neutral charge, so it is either positively or negatively charged.

Melting/freezing point: the temperature needs to be reached for a substance to melt and/or freeze.

Background and Applications:

Salty Ice Cream?

Salt is a key ingredient in making our home-made ice cream. There is an important reaction between salt and ice that helps make the ice cream. It is important to understand that salt is made out of two chemicals: sodium (Na) and chlorine (Cl). When salt is added to ice, the salt particles split apart into separate sodium and chlorine ions. An ion is an atom that has either gained or lost an electron and no longer has a neutral charge, so it is either positively or negatively charged. The salt is now made up of positive (meaning they have lost an electron) sodium ions and negative (meaning they have gained an electron) chlorine atoms. Once the ions are formed, they interfere with the intermolecular bonds of ice. Ice has a very special structure: the molecules are arranged in a rigid crystal. Once the salt ions interfere with the bonds, they make it more difficult for the ice molecules to bond together. This lowers the freezing/melting temperature, so the ice will melt even if it is below 32°F. This is also why salt is sprinkled on ice on the roads during winter. Now you may be wondering, "why would we want the ice to melt?" Well, in the context of making ice cream, the salt provides a temperature cold enough to surround the ice cream (which has a lower freezing point than water) so that the ice cream can thicken and freeze before the ice around the ice cream melts entirely.

Figure 2.9-1 Negatively charged chlorine ions and positively charged sodium ions that salt separates into once it is dissolved into water, Source: https://www.baristahustle.com/wp-content/uploads/2019/03/Table-salt-dissociate.png

Melting/Freezing points:

The melting/freezing point describes the temperature needed to be reached for a substance to melt and/or freeze. When a substance reaches a temperature below the melting/freezing point, it will freeze, changing from a liquid to solid. When a substance reaches a temperature above the melting/freezing point, it will melt, changing from a solid to liquid. Therefore, the terms melting point and freezing point can be used interchangeably. As we all know, the melting/freezing point of water is 32° Fahrenheit or 0° Celsius. Water actually has an unusually high melting/freezing point compared to other compounds. This is because of the hydrogen bonds that exist between water molecules, which require more energy to break. This special fact about water is actually crucial for life to exist on the earth. The higher melting/freezing point allows water to exist as a liquid over a large range of temperatures and support life.

Many factors can affect the melting point of a substance. These can include the addition of impurities, changes in pressure, and changes in forces within a molecule. The addition of impurities will lower a substance's melting point. This is because of how the impurities affect the structure of a compound. For example, many solids have molecules arranged in a very organized, crystalline lattice. This looks like the figure to the left. The addition of impurities will cause defects in this well-organized structure and allow the molecules to slide around. This makes it easier for them to overcome the forces holding them together. Therefore, it takes less energy for the substance to be melted, and the melting point is lowered.

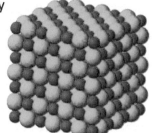

Figure 2.9-2 Crystal structure of molecules. Source: Wikipedia

Chapter 3 Physics

3.1 Orbeez by Claire Long

Materials:
1. Orbeez, preferably clear
2. Water (1 cup per ~100 orbeez)
3. Optional: Excess beaker with water filled around ½-¾ full.

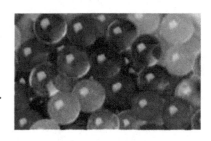

Procedures:
1. Pour water into orbeez. Observe the orbeez.
2. Wait for 4-6 hours, or leave them to soak overnight.
3. Once the orbeez have finished soaking, put them into a separate container! What changed in the size of the orbeez?
4. Optional: if you have clear orbeez, put them in a container or into the beaker with clear water and see what happens! What happened when you put them into the water?

Safety:
Don't eat the orbeez!

Experiment Questions:
1. What are orbeez made out of?
2. What causes the change in the orbeez?
3. Why do orbeez expand?
4. What is a superabsorbent polymer, and how does it work?
5. What is refraction?

Background and Applications:
Orbeez are made from superabsorbent polymers, known as SAP. Polymers are large chains of repeated molecules that are connected together. Polymers make up many things, such as cotton, plastics, and everything inside of the human body. That's right—you are made of polymers! Superabsorbent polymers are unique since they are able to absorb water extremely effectively. These polymers have chains of molecules in crossing structures that are extremely attracted to water. When SAPs come into contact with the water, these molecules attract water molecules, causing the polymer chain to get bigger as the spaces between the chains fill up with water

molecules. The polymers that make up SAP are made up of **hydrophilic,** or water attracting groups. These groups are commonly carboxylate ($-COO^-$) or sulfonate ($-SO_3^-$) groups. Both of these groups are polar, which means it is unequally charged. This results in it having the ability to share electrons with water and form hydrogen bonds with it. An example of one of these polymers is Sodium Polyacrylate. The Polymer contains carbonyl (COOH) and sodium (Na) groups. When it is in a liquid such as water, the sodium detaches from the carboxyl group, which produces a carboxyl (COO-) and sodium (Na+). Due to the water, the two groups are separated, expanding the chain. This causes more unreacted COOH and Na inside of the polymer to react, continuing the growth of the polymer. The crosslinks between these groups then are able to trap water, which gives the polymers their ability to retain water.

 Orbeez use osmosis, the diffusion of water across a membrane, to absorb water. The orbeez continue to attract water and uses it to fill up spaces in the polymer until all the spaces between the polymers are filled, in which the polymer is at full capacity. Afterwards, you can dunk the orbeez into water! What happens is that the beads completely disappear. Why does this happen? This is due to **refraction**. Refraction is a term in physics that refers to the bending and change in velocity of light as light moves from one substance to another. Different substances have different **refractive indexes**, or how the light bends in a substance. This is due to each substance having unique properties, such as mass, which change how the light is bent. Orbeez are 99% water, so the refractive index of water and orbeez is essentially the same. Outside of the water, orbeez still reflect and refract light, making it visible. Inside of the water, when light goes into the container, it has very low refraction and simply goes straight through, making the orbeez transparent and invisible inside of the water.

Figure 3.1-1 Superabsorbent polymers, upon contact with water, unravel as the sodium ions detach and the chain becomes negative.

3.2 Density Rainbow by Claire Long

Materials:

1. Five test tubes/containers
2. Warm water
3. Food coloring
4. Tablespoon
5. Sugar

Procedures:

1. Line up all of the glasses in a row and label them 1 tbsp, 2 tbsps, 3 tbsps, or 4 tbsps, leaving one blank.
2. Add 1 tbsp of sugar to the first glass, 2 tbsps to the 2nd glass, 3 tbsps to the third glass, 4 tbsps to the 4th glass, and leave the 5th glass empty. We will add 3 tbsps of warm water to each glass **except** the 5th one.
3. Add different food coloring to each glass. If you want to go rainbow order, we suggest adding red to 4 tbsps, orange to 3 tbsps, yellow to 2 tbsps, and so on.
4. Mix the food coloring well.
5. Using the sugar solution labeled "4 tbsps" dump the sugar solution into the empty container so that it takes up ¼ of the container.
6. Using a spoon, carefully layer the sugar solution that is labeled "3 tbsps" into the sugar solution until it takes up ½ of the container.
7. Repeat the 6th step for the 2 tbsps and 1 tbsp (1 tbsp goes on top), add ¼ of the test tube for each solution. Your test tube should be close to full.
8. Voila! Your rainbow test tube is complete! Now, shake up your test tube. What do you predict will happen?

Safety:

Do not spill or break the test tube!
Be careful with the warm water!

Experiment Questions:

1. Why do some items float while others sink?
2. Why does dissolving more sugar in water make that water denser?

3. When you try to mix a solution that is made of layers with different densities, what will happen?
4. What is the difference between a density tower with layers made of the same mixture as compared to ones made with different mixtures?
5. Why is the density rainbow able to mix?

Background and Applications:

Why do some objects float while others sink? Why is it that some objects/fluids are lighter than others? Well, it's because of density.

Density is the number of particles in a given space, or the measure of an object's mass per unit volume. As we added sugar to the water, the density of the water increased since there are now more sugar molecules inside of the water. In this example, given this logic, the water with four tablespoons of sugar would be the densest, followed by the three-tablespoon sugar water mixture, and so on and so forth. The denser this water is, the more it sinks, since there are more sugar molecules in the denser water, giving it more mass per unit volume than lower concentration sugar solutions. This makes the denser liquid sink to the bottom, and the least dense liquid rise to the top. This ordering of density is called **density stratification**.

However, when you mix this sugar solution with a spoon or shake it up, all the colors mix and do not separate. This is because we used different concentrations of a single molecule: sugar, or otherwise known as sucrose. Mixing this actually combines the four concentrations of sugar water into one **homogeneous mixture**, a mixture in which all particles are evenly distributed. Stirring causes the molecules to be pushed together and mixes the solution, evenly distributing the sugar in the water, creating a single solution.

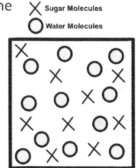

Figure 3.2-1 The homogeneous mixture of sugar and water

3.3 Air Cannon by Jessica Fan

Materials:

1. 1 Large plastic cup
2. 1 Balloon
3. Tape or rubber bands
4. 1 Drill
5. 1 Small candle
6. 1 Meter stick/measuring tape

Background and Applications:

1. *Choose the diameter of the cup's hole. A typical size ranges from ¾ to 1.5 inches. Feel free to experiment with different sizes to determine which one allows the air to travel the farthest.*
2. *Drill a hole into the bottom of the cup, directly in the center. Use the size that you decided upon in the previous step.*
3. *Place the plastic or the balloon over the large opening of the cup. Make sure it's airtight. If you are using a balloon, tie a knot in the back of the balloon, and cup the tip open OR cut off the neck of the balloon. Try blowing up the balloon to stretch it out.*
4. *Wrap a rubber band around the plastic or tape the balloon to the cup. Make sure it's airtight!*
5. *Pull back on the knot of the balloon and release it to launch (if tied) or snap your finger on the plastic membrane (for balloon and plastic wrap).*

Additional step: You can light a small candle and measure from what distance you can blow it out. Try using more force or less force to see how it affects the distance the air blows.

Safety:

Be careful with the candle, and make sure to have adult supervision.
Use the drill with adult supervision.

Experiment Questions

1. How do you calculate air speed?
2. What is associated with speed? Pressure, temperature?
3. How does releasing the balloon cause air to be pushed out of the cup?
4. How do different materials work differently in this vortex?
5. Can you find what the optimal material to use is?

Background and Applications:

The air cannon propels a vortex of air outward from the cup. A vortex is a spinning flow of fluid or gas, taking the form of a donut shape. This shape occurs because the air from the center of the hole is traveling faster than the air outside of the hole. As you pull back the balloon handle, the air is drawn into the cup chamber. Releasing causes the pressurized air to exit through the hole, directed towards a target. This forms an invisible ring of air, driven by Bernoulli's Principle.

Bernoulli's Principle states that fast-moving fluid experiences low pressure; conversely, slow-moving fluid experiences high pressure. In the air cannon, the rapid expulsion of air creates a stream of fast-moving, low-pressure air surrounded by high-pressure air, directing the airflow straight. The pressure differences cause a curling effect where air curls around the low-pressure center, forming a ring.

Adjusting the size of the hole affects the air cannon's performance: a smaller hole launches the air farther and stronger, whereas a larger hole produces a wider but weaker airflow.

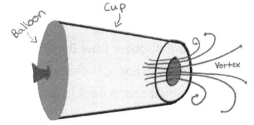

Figure 3.3-1 Air cannon

For Higher Grades:

Now, let's get a little more in depth about Bernoulli's equation.

$$P + \tfrac{1}{2}\rho v^2 + \rho g h = \text{constant}$$

His equation is split into three main parts.

Pressure (P): The pressure is the push that the fluid feels, such as the force you exert on a blown-up balloon when you press down on it.

Kinetic energy ($\frac{1}{2}\rho v^2$): This section is split into Density (ρ) and Speed (v). Density describes how concentrated the molecules are in a certain amount of space, while speed describes how fast the fluid moves. Remember, when talking about Bernoulli's principle, air is considered a fluid too, not just liquids! In its entirety, this section describes how, the faster the fluid travels, the more energy it has.

Potential energy (ρgh): This section divided into Gravity (g) and Height (h). Gravity describes the force pulling down on the fluid, while height describes how far above a certain point the fluid is. This part of the equation describes how, the higher the fluid is, the more energy it has.

We need to first think about the Law of Conservation of Energy, describing how energy can neither be created nor destroyed, but only changed. In our instance, that means trying to keep the energy in our system (whatever the equation is being applied to) the same. If the fluid is moving faster, then the pressure would have to go down in order to keep the energy the same; vice versa, if the fluid goes slower, there would be higher pressure. This balancing of energy applies to the height section of the equation as well. When the fluid changes height, its pressure and velocity may need to change to keep the energy of the system the same.

In summary, Bernoulli's Equation describes the relationship between the speed, height, and pressure of a fluid. When one changes, the other need to as well in order to maintain equilibrium (a balanced state).

Finally, we will discover how Bernoulli's principle applies to the Air Cannon experiment. When we pull back on the balloon, it creates less space for the air, creating high pressure. When we release the balloon and it snaps back in place, it rapidly pushes air out of the cup.

As air is forced through the small opening of the cup that you drilled into the base, the air speeds up. However, according to Bernoulli's equation, when the fluid speeds up, the pressure goes down. The movement of the air also creates a vortex as it is forced out of the cup, allowing the air to maintain a ring shape as it travels forward.

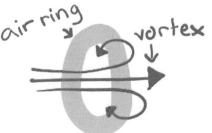

The image shown depicts how the air rings are formed after the air is pushed out of the cup.

Figure 3.3-2 Air vortex

3.4 Capillary Action by Claire Long

Materials:

1. 1 white flower (daisy or chrysanthemum)
2. 1 cup
3. Warm water
4. Food coloring of your choice

Procedures:

1. Fill the container (cup) with warm water.
2. OPTIONAL: Use a knife to slice the bottom of the stems at a 45-degree angle. Scissors are not recommended, as they can end up crushing the flower's stem.
3. Add drops of food coloring of whatever color you want and stir.
4. Stick the flower into the container.
5. Wait for the color to travel (Around a day, wait for 3-5 days for best results).

Safety:

Don't consume the flower!

If you are to cut the flower stems using a knife, do so with adult supervision.

Experiment Questions:

1. What is capillary action?
2. What is adhesion, cohesion, and surface tension?
3. Why does capillary action occur?
4. What is the difference between capillary action and chromatography process?
5. How do plants use capillary action?

Background and Applications:

This experiment helps you understand the processes of capillary action, which is the movement of water through a material, or in our case, a tube, without or against the assistance of gravity.

Plants are able to use capillary action to distribute water throughout the plant. To do so, plants use xylem to transport water up from their roots. However, this ability to transport water depends on the inherent properties of water.

Water has three properties: **adhesion**, **cohesion**, and **surface tension**. **Adhesion** is the property of water that allows it to stick to other things, while **cohesion** is the property of water that allows water to stick to itself. **Surface tension** is an extension of cohesion, and it is defined as the property that causes the surface of water to resist outside forces and remain intact, keeping water molecules together.

Capillary action occurs when the water's force of adhesion to the xylem is stronger than that of cohesion (water's adhesion to itself). Then, thanks to transpiration, the loss of water vapor from the leaves, as water evaporates from pores called stomata on the leaf surface, a suction force is created. As one molecule of water sticks to the walls and moves up the stem thanks to the suction force, cohesion ensures that the other molecules of water follow and pull each other along. Surface tension ensures the molecules are held together throughout this process, and allows water to be continuously pulled up the xylem.

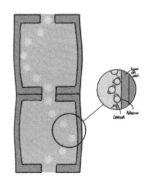

Figure 3.4-1 Water attaching to a xylem cell wall using adhesion and cohesion

Capillary action can occur in any narrow cylindrical tube, as long as there is intermolecular attraction between water molecules and adhesion between the walls of the tube and the liquid. Capillary action can also occur in spaces with high porosity: the amount of void space in a material. Porous materials, such as soils, rock, and sponges, have more space for liquid to move through. Water adheres to the pores, and the cohesive force of water makes water stick together to create a continuous column of water. The surface tension of water then keeps the surface of the water together in the shape of a meniscus, and the forces combine to allow water to climb through the narrow spaces, or pores, of the material, called capillary rise.

Capillary pressure also has an impact: capillary pressure is the pressure difference caused by the balance of cohesive and adhesive forces against gravity. As capillary pressure is generated, the liquid moves upwards against gravity until the two forces balance each other out. This capillary pressure can be calculated as $P_c = \frac{2T}{R}$, where P_c is the capillary pressure, T is the surface tension, and R is the radius of curvature of the meniscus, the curve in the surface of a fluid when it touches another material.

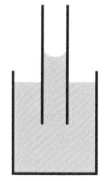

The overall capillary rise equation is $h = \frac{2T\cos\theta}{\rho g r}$ where h is the capillary rise, T is the surface tension, θ is the contact angle between the liquid and the tube, ρ is the density of the liquid, g is gravitational acceleration (usually 9.81 m/s), and r is the radius of the tube.

Adhesion > Cohesion

Figure 3.4-2 Capillary action

3.5 Instant Ice by Jessica Fan

Materials:
1. 4 plastic water bottles
2. Fridge with a freezer
3. 4 ice cubes
4. Plate/bowl (something to hold the ice cube in and catch the water)

Procedures:
1. Place the 4 water bottles in the freezer.
    a. The first bottle stays in the freezer for 30 minutes.
    b. The second bottle stays in the freezer for one hour.
    c. The third bottle stays in the freezer for 1 hour and 30 minutes.
    d. The fourth bottle stays in the freezer for 2 hours.
Make sure that the water inside of the bottles doesn't freeze up. If you need to take it out before the 2-hour mark, that's totally fine. Just try to have the bottle in the freezer for as long as possible.
2. Carefully remove the water bottles from the freezer, and DO NOT SHAKE THEM.
3. Place an ice cube on top of the pouring surface.
4. Uncap the bottled water.
5. Slowly pour the water onto the ice. Do this four times, with each bottle on a different ice cube.
6. Does it form an ice tower?

Safety:
Be sure to clean up to prevent injuries from sliding on water.
Be careful around frozen objects.

Experiment questions:
1. How does increasing the time the water spends in your freezer correlate with the height of the ice tower?
2. Did a longer time or shorter time work better?
3. Why does this experiment not work if the water is already half frozen?
4. Why does this experiment not work if the water is not cold enough?
5. Does the type of fridge affect the freezing time?

Background and Applications:

Have you ever wondered how crystals form? What about the intricate patterns of snowflakes? The answer lies within nucleation.

Now, you might be wondering: what does a nucleus have to do with ice cubes and crystals? Are we talking about an atom or a cell's DNA? Not quite! The nucleation we're exploring today is the process by which liquid transitions into a crystalline solid.

The formation of a crystal involves two primary steps: nucleation and growth. Let's start with nucleation.

Nucleation begins with a liquid, such as nearly frozen water in our experiment, that has yet to crystallize. Nucleation is the process where ions, atoms, or molecules arrange themselves into a pattern characteristic of a crystalline solid. This can happen spontaneously or whenever an impurity, such as dust, is introduced to the system.

Once nucleation occurs, the initial cluster of arranged molecules is a seed from which the rest of the crystal will grow.

Here is an example to illustrate the process of nucleation: in the water cycle, during condensation, the evaporated water begins to change back into its liquid form. At this stage, dust particles may be introduced from the atmosphere and combine with the water molecules, acting as nucleation sites. If the temperature is below freezing, the water droplets will freeze around these dust particles, forming ice crystals.

The second stage is growth. Once a seed crystal forms, the surrounding liquid can continue to crystallize around it. In the case of snowflakes, water droplets freeze layer by layer, forming the beautiful six-sided pattern characteristic of snowflakes.

You might be wondering: what does this have to do with the experiment we just did? Well, the same process is at play here. Instead of a dust particle, the ice acts as the seed. When the cold, nearly frozen water comes in contact with the ice, it begins to crystallize, forming the beautiful ice tower like magic!

There are many different types of nucleation, as depicted in the image shown. This diagram illustrates the types of freezing within the atmosphere, with explanations provided beneath.

Homogenous freezing: Occurs when the temperatures are low enough for molecules to freeze without any sort of nucleation site, such as dust or other small particles.

Deposition nucleation: Occurs when particles in the atmosphere are present, and water vapor directly crystallizes onto the particle without the in-between liquid stage. This is why it is called deposition, because deposition is the term for a phase change where gas freezes directly into a solid form.

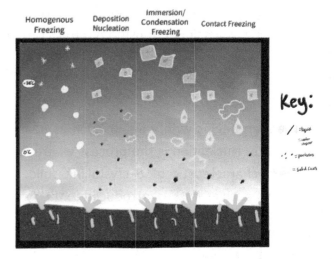

Figure 3.5-1 Ice nucleation mechanism in atmosphere

Immersion/Condensation Freezing: The normal kind of freezing that we think of in the atmosphere, where water condenses around a particle and then freezes when the temperature is low enough.

Contact freezing: Occurs when supercooled water in the atmosphere encounters a nucleation site, such as dust, causing instant freezing into ice crystals. This phenomenon mirrors what we observe in our instant ice experiment, where supercooled water freezes upon encountering a nucleation point.

Next time you gaze up at the sky filled with snowflakes, you will understand how they are formed and which processes they went through to create the intricately designed weather phenomena.

3.6 Oobleck by Claire Long

Materials:

1. 1 Cup cornstarch
2. ¾ Cup water
3. Bowl
4. Food coloring
5. Popsicle stick or something to stir with

Procedures:

1. Mix 1 cup cornstarch, ¾ cup Water, and a couple of drops of food coloring in a bowl.
2. Hardens with Physical Trauma! Try to punch it and hold it!

Safety:

Be careful when mixing!

Please do not eat the Oobleck.

Do note that Oobleck can be quite messy: do it in a place where it can be cleaned up easily.

Experiment Questions:

1. What is the difference between a Newtonian fluid and a non-Newtonian fluid?
2. Is Oobleck an example of a non-Newtonian or Newtonian fluid?
3. What are examples of Newtonian and non-Newtonian fluids?
4. Why does Oobleck get harder to manipulate when you add more force?
5. Why does the Oobleck seem to be liquid when not moved?

Background and Applications:

This experiment is an example of a **non-Newtonian fluid.** What is a non-Newtonian fluid? To answer that, we'll have to first define what a Newtonian fluid is.

A **Newtonian fluid** is a fluid where the viscosity (the thickness) remains the same no matter how much shear force (otherwise known as stress) you apply. Stress is the force that is applied to an object: this can be something like a push or a pull. When you push, pull, or otherwise act on a Newtonian fluid, the fluid will not suddenly get thicker or thinner: the viscosity stays the same. Examples of Newtonian fluids include water, alcohol, and air.

A **non-Newtonian fluid** is a fluid that changes viscosity depending on the stress applied to it. The viscosity can either increase or decrease, depending on the fluid. For Oobleck, the viscosity increases as you apply more force, meaning that it gets thicker and harder to manipulate. These kinds of fluids are called **dilatants** or **shear thickening,** and other examples include quicksand and silly putty. However, for some fluids, such as ketchup, the viscosity actually decreases. For example, if you shake ketchup in a bottle, the ketchup thins out. These fluids are called **pseudoplastics** or **shear thinning**, and common examples include nail polish, lava, and whipped cream.

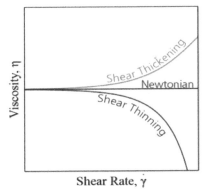

Figure 3.6-1 A diagram of shear rates compared to velocity for shear thickening, Newtonian, or shear thinning fluids. Source: www.rheosense.com

Either way, in non-Newtonian fluids, suddenly applying a force causes changes in viscosity. In dilatants, the large molecules in the fluid lock up, since they cannot move apart in time in reaction to the force, causing them to act like a solid. In pseudoplastics, the pressure causes the molecules to reorganize, making them more liquid-like.

3.7 Balloon Laminar Flow by Carter Feng

Materials:

1. 1 balloon
2. A pair of sharp scissors
3. 1 roll of tape, preferably electrical tape or any other sticky and thick tape
4. Sink

Procedures:

1. Put the balloon mouth onto the sink head and turn on the water.
2. Fill the balloon until it is about the size of your head.
3. Take the balloon off the sink head.
4. Squeeze the balloon and get rid of the air inside the balloon.
5. Tie up the balloon.
6. Put the tape in a tic tac toe shape on the balloon with a space about 1 cm x 1 cm in the middle.
7. Use the scissors and poke a hole in the space in the middle of the tape.
8. The water coming out should appear as if it is "paused".

Safety:

Be careful with the scissors and make sure you don't poke yourself.

Experiment Questions:

1. In what other ways do you think you can exhibit laminar flow?
2. What is laminar flow characterized with?
3. What is intermittent flow?
4. How could laminar flow be used in real life?
5. What is the Reynolds number and how is it used to describe whether a flow is laminar, turbulent, or intermittent?

Background and Applications:

This experiment helps you understand what laminar flow is and in what circumstances it can happen. Laminar flow is when a fluid's particles flow smoothly in paths like layers, with each layer moving smoothly past the adjacent layers with little or no mixing.

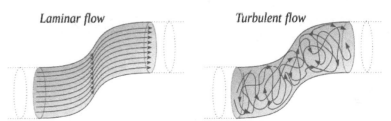

Figure 3.7-1 Laminar and turbulent flow. Source: fossilconsulting.com

Turbulent flow is when the water is chaotic, causing the "layers" to mix. An example of turbulent flow is if you poked a hole in a balloon filled with water without taking out the air or adding tape. However, when you take out the air in the balloon, it creates an environment in which laminar flow can happen. The tic-tac-toe pattern made of tape controls the water and helps stabilize the flow.

Turbulent flow is also categorized when the Reynolds number is over 2900. Reynolds number is the ratio of inertial forces to viscous forces. It is a dimensionless number that is used to categorize fluid systems. The Reynolds number is defined as $N_{Re} = \frac{\rho v d}{\mu}$, where ρ = density, v = velocity, d = diameter, and μ = viscosity. When the Reynolds number is less than 2300, it is assumed as laminar flow, and if it is above 2900 it is turbulent. When it is below 2900 but not above 2300, it is assumed as intermittent flow or transitional flow. Intermittent flow will form continuous turbulent flow but only at a very long distance from the inlet. The flow between the continuous turbulent and the inlet transitions from laminar to turbulent and then back to laminar at irregular intervals. This happens because of the different speeds in different areas of the pipes that depend on factors such as pipe roughness.

In real life, laminar flow is mostly used in cases with liquids or gases flowing through pipes/ducts. This flow requires less energy and is more consistent, allowing it to be predicted more easily.

Supervision by an adult is recommended.

3.8 Balloon Competition by Jessica Fan

Materials:

1. *Straw*
2. *Tape*
3. *1 Balloon*
4. *String*

Procedures:

1. *Blow up your balloon and pinch the end. DO NOT TIE.*
2. *Cut your straw to about 2 in. in length.*
3. *Tape the straw to your blown-up balloon.*
4. *Slide the straw onto the string (it's ok if your balloon shrinks a little while doing this).*
5. *Blow up the balloon.*
6. *Let go!*
7. *Now, try adding weight to your balloon (perhaps taping on a penny).*
8. *Repeat by adding more and more weight.*
9. *Try blowing up the balloon with more or less air.*

Safety:

Eye protection is recommended.

Experiment Questions

1. What happens if you add weight to the balloon?
2. What happens if you blow more air or less air? How far does the balloon go?
3. Why does the balloon slowly deflate as it moves forward?
4. What happens if there is a hole in the balloon that's not the opening? Does it impede the balloon's speed?
5. Does changing the type of balloon change the speed? Why or why not?

Background and Applications:

Why does the balloon blast away? When you blow up a ballon, the rubber stretches, storing elastic energy, which is a form of potential energy. Potential energy is stored energy, and elastic energy occurs when an object is stretched or compressed. The air wants to be pushed out of the balloon because of pressure differences inside and outside of the balloon, but the air is surrounded by rubber and can't escape. When you release the balloon, the rubber contracts rapidly, forcing the air out through the small opening when you let it go.

This phenomenon exemplifies Newton's third law of motion: "For every action (force) in nature, there is an equal and opposite reaction." This means that when one object puts a force on another, the other object puts an equal and opposite force back.

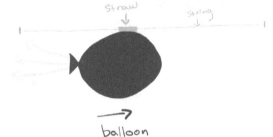

Figure 3.8-1 Ballon competition illustration

Newton's third law can be illustrated by imagining an astronaut in space. If the astronaut drifts away from the space station and throws a wrench in the opposite direction, the force applied to throw the wrench propels the astronaut back toward the station. Since the astronaut puts a force on the wrench, it puts an equal force on the astronaut, sending him back toward the space station. The same concept is shown with a balloon rocket, where the air escapes one way and propels the balloon in the opposite direction. You can experiment with how much you blow up the balloon to find the size that goes the farthest.

Although you may not realize it, Newton's third law is at work all around us, from normal, everyday activities like walking to rockets being launched off into space. Understanding this law is crucial because it explains how forces interact and move objects in our world.

3.9 Slinky 'EM by Jessica Fan

Materials:

1. Slinky

Procedures:

1. Two people hold each end of the slinky.
2. One person pushes the end of the slinky directly toward the other person. (This is a primary wave).
3. Next, that person moves their end up and down. (This shows a secondary wave).

Safety:

Please don't pull the slinky fast or twist it, as slinkies get tangles extremely easily.

Experiment Questions

1. How does the motion of a slinky model show different kind of waves?
2. Are there other types of waves besides those two?
3. What are some limitations to the slinky model? Why can't we use real waves instead?
4. Why do you think the slinky model is effective compared to other models?
5. Can other types of waves be modeled using a slinky? Why or why not?

Background and Applications:

Have you ever wondered how earthquakes happen? Why can the simple movement of the earth cause so much damage? The answer lies in waves!

No, not ocean waves. The waves we're discussing today are seismic waves.

There are two main types of seismic waves: body waves and surface waves. The kind of waves that we demonstrated with the experiment were body waves.

When earthquakes first happen, primary waves causing the shaking. The waves, also dubbed P-waves, travel quickly through both liquid and solid, including earth's molten core. Primary waves cause particles to move in the same direction as the wave's propagation, similar to compressing and stretching a slinky. This motion is illustrated here:

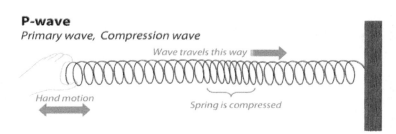

Figure 3.9-1 P-wave

The places where the lines/particles are compressed is called the compression, while the place where the lines/particles are spread apart is called rarefaction. In an earthquake, the p-wave is the sharp jolt or thud that you feel at first.

Following primary waves are secondary waves, or S-waves, which travel more slowly because they are unable to travel through liquids like P-waves can. These waves travel when the particle is moving perpendicular to the direction of propagation, as seen in figure 3.9-2:

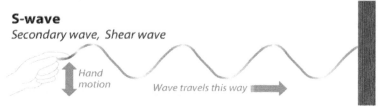

Figure 3.9-2 S-wave

This causes an up or down or side-to side motion in an earthquake.

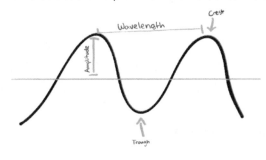

Figure 3.9-3 Transverse wave diagram

As you can see, the waves are moving forward while the particles are moving up and down. These waves are usually more destructive than P-waves, because while a P-wave is a quick jolt, the S-waves undulate up and down or side to side, causing a lot of damage to buildings. Now you understand how body waves work in an earthquake!

3.10   Ice Ice Baby by Emma Zeng

Materials:

1. Sugar
2. Salt
3. Water
4. Cups
5. Food coloring

Procedures:

1. Put equal amounts of water into three cups.
2. In one cup, add salt, and stir.
3. In another cup, add sugar, and stir.
4. Leave the third cup with just water.
5. Put ice cubes of equal weight and surface area inside and time how long they take to melt.
6. Add a drop of food coloring into each cup and see how the color moves.

Safety:

Don't spill the water! If you do accidentally spill it, make sure to clean it up right away.
Don't eat any of the experiment materials.

Experiment Questions

1. Which cup do you predict will melt the ice the fastest?
2. Which cup do you predict will melt the ice the slowest?

Background and Applications:

This experiment takes a look at convection and density in water. After completing the experiment, we should have noticed that the ice in the normal water had melted the fastest, while the ice in the salt and sugar mixture had taken much longer.

This is because the salt and sugar mixtures are much denser than normal water. As the ice melts in the salt and sugar water, the water melting from the ice cube is less dense than the mixture of salt or sugar. This water won't sink or mix in with the denser water. Instead, it sits at the surface, in its own colder water section, slowing down the melting process. Imagine it like oil sitting on water, the oil is less dense, making it hard to mix or sink below the water.

Meanwhile, normal water circulates cold and warm water, making the ice cube melt faster. Because both the ice cube and normal water solution have the same density, they mix together well, evening out the average temperature.

This can also be seen when we add a drop of food coloring to each cup. In the normal water solution, the food coloring quickly circulates and spreads through the cup, while the food coloring in the salt and sugar water sits at the top, unmoving.

normal water   salt water

Figure 3.10-1 The food coloring diluted and circulating in freshwater vs it sitting at the top in salt water.

3.11   Heat Convection by Emma Zeng

Materials:

1. Red food coloring
2. Blue food coloring
3. Warm/hot water
4. Ice cube tray
5. A clear tub filled with water

Procedures:

1. Using the blue food coloring and the ice cube tray, make frozen blue colored ice cubes.
2. Using the red food coloring, make the hot water red.
3. Place the ice on the top right of the tub, making sure not to splash
4. Dump the hot water gently into the top left of the tub
5. Make a prediction: What will happen when we add the red-hot water and blue-cold water?
6. Watch as the colors swirl - warm water rises, cool water sinks

Safety:

Be careful around hot water.

Experiment Questions:

1. Applying this to another scenario, what happens when boiling water over a flame?

2. Can we see convection currents with warm and cold air?

Background and Applications:

Through this experiment, we can see convection currents in water. These cycles are driven by the different temperatures in heat and their proximity to the heat source. Using the colored water, we can see that warm water rises and cold-water sinks. These convection currents happen all around us. A prime example is the convection currents in the earth's mantle layer, right under the crust. As the mantle closer to the earth's core warms up, it rises. On the other hand, as the mantle closer to the crust cools, it sinks. This creates circular motions within the mantle that cause the movement of tectonic plates.

So, what causes the warm water to rise and the cold water to sink? This is a matter of density. In cold water, the molecules are all close together and moving at a slow speed. This increases the density of cold water, allowing it to sink below the room temperature water. On the other hand, the molecules in warm water move faster and are more spread out. This gives warm water a less dense makeup that allows it to rise and rest on top of the room-temperature water.

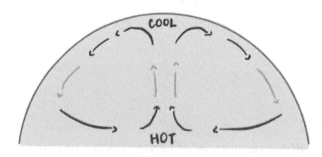

Figure 3.11-1 The cold and hot currents in the Earth's mantle as it gets closer and further from the core.

Supervision by an adult is recommended.

3.12 Can Implosion by Emma Zeng

Don't do it without an adult supervisor!

Materials:

1. Soda can
2. Frying pan
3. Water
4. Bowls
5. Ice
6. Water

Procedures:

1. Put a small amount of water into two soda cans (about ½ inch).
2. Set the cans in the pan and heat them until the water boils.
3. While the water is boiling, get two bowls of ice water ready.
4. When steam starts coming out of the cans, use tongs and quickly turn a can upside down into the ice water.
5. Observe what happens!

Safety:

Don't scald yourself! Make sure to boil the water under adult supervision.
Wear safety goggles throughout the experiment.
The aluminum can be sharp, so be careful!

Experiment Questions

1. What do you predict is going to happen, and why?
2. Why doesn't this experiment work with warm water?

Background and Applications:

To understand this experiment, we first need to understand air pressure. When the molecules of our atmosphere move, they hit boundaries, exerting a small amount of force. When we look at all the molecules and their combined force, we see their air pressure. This air pressure can be increased or decreased in two ways. One is the addition or subtraction of air molecules. More molecules would mean more particles hitting boundaries, increasing the sum force. A second way would be the addition and subtraction of heat. Heated molecules increase their energy, increasing their velocity and speed as they hit boundaries and effectively changing air pressure.

This experiment examines thermodynamics and air pressure. When the can is heated, the water inside boils and turns into water vapor. The vapor then pushes all the air out of the can, taking up all the space inside. Then, when turned upside down and plunged into cold water, the sudden change of temperature turns the water vapor from a gas into a small amount of water. Since plunging it into water blocks the only open hole, an exit and an entrance for air, the sudden decrease of air within the can (because gas → water) leads to an imbalance of air pressure within the can. This sudden change uses the two methods we reviewed earlier: the subtraction of air molecules as well as the subtraction of heat. As the pressure is now lower in the can, the higher-pressure water and air surrounding pushes inward on the can. This higher-pressure force, as seen, crushes the can.

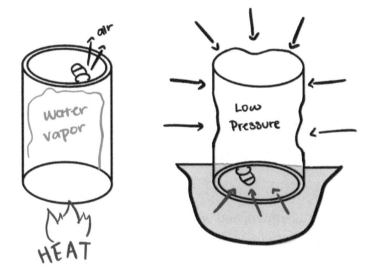

Figure 3.12-1 Can implosion, the can when placed in the cold water. The arrows represent the higher-pressured water and air pushing on the can.

3.13   Egg in a Bottle by Carter Feng

Materials:

1. 1 peeled hard-boiled egg
2. 1 glass bottle/jar with the head slightly smaller than the egg
3. Lighter/matches
4. 1 sheet of paper

A slip of burning paper is put into the bottle

An egg is put on the top of the bottle

The egg is sucked into the bottle

Procedures:

1. Crumple the sheet of paper into a ball and light it with the lighter/match.
2. Put it in the bottle.
3. Put the egg on the bottle's mouth.

Safety:

Don't burn yourself.
Use gloves for lighting the paper.
Wear goggles for eye protection.

Experiment Questions:

1. What is the Ideal Gas Law?
2. The egg could be more easily put into the bottle by doing what?
3. Which is more dense, hot or cold air?

Background and Applications:

During this experiment, the burning paper heats the air. As the molecules heat and move faster, the hot air expands. The egg is placed on top of the bottle, creating a seal that keeps the air from escaping. Once the paper burns out, the air inside the bottle begins to cool. As the air inside the bottle cools, it contracts, creating a partial vacuum. Outside air pressure is now greater than the pressure inside the bottle. The pressure difference between the inside and outside of the bottle causes the egg to be pushed into the bottle.

This experiment helps you understand the Ideal Gas Law (PV = nRT, where P = the pressure of the gas, V = the volume taken up by the gas, T = the temperature of the gas, R = the gas constant, and n = the amount of moles of the gas).

When explaining the Ideal Gas Law, you must first learn about what an ideal gas is. Ideal gas molecules do not attract or repel each other and take up no volume. The gas takes up volume but the molecules are approximated as particles that have no volume. There are no gases exactly ideal but there are many that are close which makes the concept a very useful approximation. However, if the pressure of the gas is too large, or the temperature is too low there can be significant deviations from the ideal gas law.

From the Ideal Gas Law, you can know that when you heat up the air, the pressure must increase because the volume, gas constant, and the amount of moles stay constant.

The Ideal Gas Law was derived from Boyle's law and the other simple gas laws. Boyle's law is the relationship between the pressure and volume of a confined gas. Boyle's Law is $P_1 V_1 = P_2 V_2$, where P_1 = first pressure, V_1 = first volume, P_2 = second pressure, and V_2 = second volume. In simple terms, if you squeeze a gas into a smaller space, its pressure goes up, and if you let it expand into a larger space, its pressure goes down.

3.14 Centripetal Force by Carter Feng

Materials:

1. 4 ft of rope
2. 1 thin square wooden board
3. 1 drill
4. 1 plastic cup
5. Water
6. 1 square sheet of rubber (smaller than the board)

Procedures:

1. Drill a hole in each corner so that the rope can fit through.
2. Glue the thin sheet of rubber to one side of the board.
3. Tie the 4 ropes together at one end.
4. Pass each rope through the corresponding hole and tie the ends.
5. Place the plastic cup on the rubber.
6. Pour water into the cup until it is about halfway full.
7. Slowly begin swinging the board and swing it in a complete circle when you are ready.
8. Slowly bring the board to a stop.

Safety:

Be careful and using the drill under adult supervision

Experiment Questions:

1. There is another force called centrifugal force. This force is the one you feel when you spin the ball, as if it is pulling outwards. What is this caused by?
2. A car on a circular racetrack cannot stay on the track without what?
3. A method used to make it easier for the car to stay on the track is to bank the road by some angle. How does this make it easier?
4. Increasing the length of the radius by 1 foot decreases the force needed by what factor?
5. Increasing the speed of the board by two times increases the force by what?

Background and Applications:

This experiment helps you understand Newton's First Law of Motion, which is that objects in motion tend to remain in motion unless acted upon by an external force.

For the cup to remain seated on the rubber during this experiment it needs to move in a tangent along the circle. The force that keeps it turning toward the center of the circle is the tension in the rope. When you spin the ball, the force applied to the ball is tangent to the circle. However, instead of flying off, the ball moves in a circle because the rope constantly pulls the ball inward, causing it to change direction.

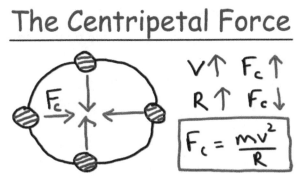

Figure 3.14-1 Centripetal force. Source: The Organic Chemistry Tutor

Even when the water is at the top of the circle, the force continues to accelerate the water downwards in the circular pattern, keeping it from spilling down on your head.

You might be thinking that gravity would cause the water to pour out when it is at the top of the arc, but this is not true. The force of gravity is 9.8 m/s^2 times the mass, which would be less than the force that keeps the ball in circular motion as long as the acceleration remains high enough.

3.15 Cartesian Driver by Carter Feng

Materials:

1. 1 water bottle
2. 1 sauce packet
3. Water

Procedures:

1. Fill the bottle to the brim with water.
2. Place the sauce packet on the water surface.
3. Screw the cap back on.
4. Squeeze the bottle and see how the packet moves down.

Safety:

Please make sure to have adult supervision.

Do not drink the water used in this activity.

Experiment Questions:

1. What happens when the bottle is squeezed?
2. What happens when the bottle is released?
3. Why must we fill the bottle completely with water?
4. What would happen if the bottle was only halfway filled?
5. Fish use this method to go up or down in water. How?

Background and Applications:

When the bottle is squeezed, the pressure inside increases. The increase of pressure compresses the air in the diver, which makes it denser. The diver becomes too heavy for the water to keep it at the top of the liquid, causing it to sink to the bottom of the bottle.

When you stop squeezing on the bottle, the pressure decreases. The air bubble in the diver expands and becomes less dense. This makes it rise to the top again.

For Higher Grades:

This experiment helps you understand density. Density describes substances based on how much mass they have in a certain amount of space they take up, the volume. Density is defined as $\rho = \frac{m}{V}$, where ρ = density, m = mass, and V = volume.

To find the relationship between pressure and density, we must first know the density equation and the Ideal Gas Law (PV = nRT, where P = the pressure of the gas, V = the volume taken up by the gas, T = the temperature of the gas, R = the gas constant, and n = the number of moles of the gas). Knowing that

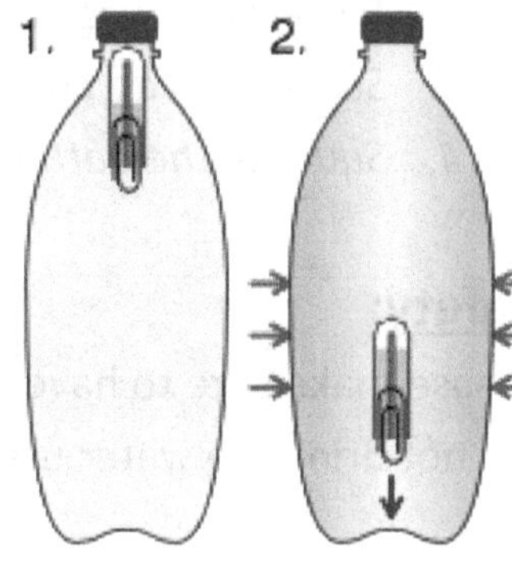

n = amount of moles

$$= \frac{m}{M} \text{ (m=mass and M = molar mass)}$$

Figure 3.15-1 Cartesian driver. Source: Wikipedia

you can get that $PV = \frac{m}{M}RT$. Plugging in the density formula gets you $P = \rho\frac{R}{M}T$. $\frac{R}{M}$ = the specific gas constant, so you end up with $P = \rho R_{specific}T$.

Basically, when you squeeze the bottle, the pressure increases which means the density of the packet increases, so it sinks. When you release the pressure, the packet decreases in density so it rises.

Chapter 4 Earth Science

4.1 Starburst Rock Cycle by Jessica Fan

Materials:

1. *2-5 Starbursts*
2. *Scissors*
3. *Microwave*
4. *Disposable, microwaveable bowl*

Procedures:

1. *Cut Starburst into pieces to simulate weathering.*
2. *Pack Starburst "sediments" into large balls.*
3. *Knead Starbursts with your hands until they fade into each other.*
4. *Put Starbursts into the microwave for 30 seconds.*
5. *Wait for Starbursts to cool and harden.*

Safety:

1. The Starbursts, as well as the bowl, may get hot after being microwaved. Be careful when handling these.
2. The Starbursts may be hard to remove from the bowl after cooling, so heat the bowl with warm water to remove the Starbursts, or throw the bowl away.

Experiment Questions:

1. Have you ever wondered where rocks come from?
2. What do you think each step of our experiment represented in the rock cycle?
3. How would the experiment have differed if you melted the rocks right after cutting them? What do you think skipping part of the rock cycle means?
4. What other materials would work for the Starburst rock cycle, or do you think Starbursts are the most effective?
5. Why do you think that, even though rocks are hard, they can still melt?

Background and Applications:

Human activities, animal interactions, and various natural processes constantly change our earth. One of these natural processes is the rock cycle, which gradually alters the earth's surface over time. Although this process has no distinct beginning or end, we can conceptualize its stages to understand how existing rocks on and within the earth undergo transformation.

The rock cycle starts off with weathering, when rocks at the surface of the earth are broken down into sediments. This can happen in three ways: chemical, physical, or biological.

Chemical weathering occurs when the individual minerals inside the rock are broken down, resulting in new minerals. For example, acid rain can dissolve rocks and create different mineral compositions.

Figure 4.1-1 Acid rain breaks down rocks

Physical weathering breaks down rocks into smaller fragments but doesn't alter their chemical composition. Agents of weathering include wind, water, and ice, which gradually wear down rocks into sediments.

Figure 4.1-2 Other weather breaks down rocks

Finally, biological weathering occurs when rocks are broken down through natural processes, such as grass roots slowly breaking apart soil or tree roots fracturing rocks. An example is illustrated below.

The process of sediments becoming sedimentary rock involves lithification. This is when the sediments compact under pressure, and their liquids are expelled, making sedimentary rock. Examples of sedimentary rocks include sandstone, breccia, and shale.

Figure 4.1-3 Examples of igneous rocks

Next, **metamorphic rocks** form through heat and pressure. There are two types of metamorphism: regional and contact. Regional metamorphism affects a broad area of crust, while contact metamorphism occurs when rocks are exposed to the heat of nearby igneous intrusions like dikes or magma chambers. Examples of metamorphic rocks include Gneiss, Serpentinite, and Quartzite.

Next, **igneous rocks** are formed when magma cool and solidifies. **Magma** results from rocks melting due to the earth's internal heat. However, when the magma erupts from a volcano, it becomes **lava**; there is no chemical difference between magma and lava, only the location difference. Igneous rocks are formed from cooled magma/lava and are classified into two types: Plutonic/Intrusive, and Volcanic/Extrusive.

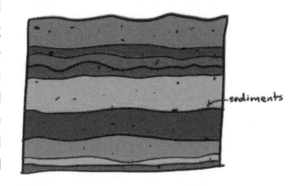

Figure 4.1-4 Sediments

Plutonic/Intrusive rocks are formed below the earth's surface when magma cools slowly. **Volcanic/Extrusive** rocks form from lava when a volcano erupts, cooling rapidly at the earth's surface. Examples of igneous rocks include Pumice, obsidian, and Basalt.

Understanding the rock cycle gives us an appreciation of earth's geological processes which constantly reshape the world we live in.

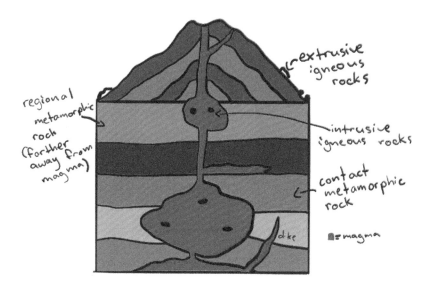

Figure 4.1-5 Examples of igneous rocks

4.2 Cloud Collage by Brian Lu

Materials:

1. Blue sheet of paper
2. 8 cotton balls
3. Glue

Procedures:

1. Cumulus clouds: Use 4 cotton balls and stick them close together on the construction paper.

   (Looks like this→)

2. Stratus clouds: Stretch 3 cotton balls to be flat and a little puffy. Glue to the bottom of your

   paper. (Looks like this →)

3. Cirrus clouds: Pull apart 1 cotton ball so it is sort of see through. Glue to the top of your

   paper. (Looks like this →)

Safety:

Do not eat the glue or choke on the clouds. Wash your hands after using glue.

Experiment Questions:

1. What are Clouds?
2. How do clouds form? Why are they so important to life on earth?
3. What is Adiabatic Cooling?

Background and Applications:

Clouds, we know them, we love them, and they give us life and water. But how do they form, and why do they look the way they do? While the reason why clouds differ significantly in appearance is beyond the scope of this book or any other entry level science book, we can talk about the general process behind the formation of clouds. Our atmosphere holds a certain amount of water per cubic meter, which is dependent on temperature. As we increase our altitude in the troposphere, our temperature will drop, decreasing the amount of water we can hold. This is important because warm(er) air rises, meaning that air masses humidified by evaporating water will rise. As the air mass rises, the air pressure drops, so the temperature drops, decreasing the amount of water the air mass can hold. When the temperature reaches a critical threshold, water molecules in the air will start condensing onto particles of dust in the air—it doesn't just clump to itself for very complex reasons.

Figure 4.2-1 Cloud formation. Source: NASA/JPL

Now, let's talk about the various types of clouds. For simplicity, we will be only modeling the basic cumulus, stratus, and cirrus clouds with our cotton balls. In our atmosphere, there are various layers, for example the troposphere, stratosphere, mesosphere, etc. All (except a few exceptions) clouds exist in the troposphere.

We have low level clouds, medium level clouds, and high clouds. Along with these altitude classifications, we have 4 main cloud forms, stratiform, cirriform, cumuliform, and cumulonimbus, each with their own distinct shapes and behavior.

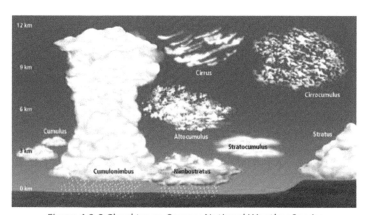

Figure 4.2-2 Cloud types. Source: National Weather Service

Stratiform clouds. also called layer clouds, are clouds that are flat in shape, and do not have any vertical expansion. These are formed in air conditions that are very stable. This cloud also has the subclassifications altostratus and cirrostratus, which denotes the middle and high-level variants respectively.

Next, the cirriform clouds, which are always very high level, exist in the high level. They are always composed of mainly ice at -25°C to -85°C. They tend to be tinted orange/red even at midday due to their high altitude and refraction of the sun's light.

We also have the cumuliform clouds, which are defined by rounded, almost cotton-ball like shapes. They form when warm, humid air rises through cooler air. These clouds do not generally produce precipitation, but when they do, it is light. This cloud also has its respective mid-level (Alto) and high level (Cirro) variants.

Now, we come to the cumulonimbus clouds, which are tall, anvil shaped clouds. These are generally referred to as thunderstorm clouds and are produced by unstable air masses and lifting forces.

Finally, we come to the nimbostratus clouds, which are large sheets of precipitation generating clouds. They don't fit cleanly into any of the categories, other than being low level clouds.

4.3 Rain in a Jar by Brian Lu

Materials:

1. Water
2. Shaving cream (1 bottle)
3. Food coloring
4. Pipette
5. Clear jar/bottle

Procedures:

1. Fill up around 4/5 of the jar/bottle with water.
2. Squeeze shaving cream onto the water in a spiral.
3. Using the pipette squeeze a solution of 4 drops of food coloring per cup of water into the shaving cream.
4. Add more if rain does not appear immediately.

Safety:

Don't drink shaving cream

Experiment Questions

1. What types of clouds generate precipitation?
2. Write out the ideal gas law, and the associated variables.
3. What are the 4 common types of precipitation?

Background and Applications:

Clouds? What are they? How do they form? And why are they so important? Clouds are just a small part of the immense field of meteorology. In this experiment, we will be discussing how clouds form, and how they relate to the rest of meteorology.

First, what are clouds? Clouds are visible masses of condensed water vapors floating in the air, otherwise referred to as an aerosol. They form when water vapor is created through the evaporation of water and rises and cools enough through adiabatic cooling (cooling due to the drop in pressure, rather than the heat escaping) and condensing onto atmospheric dust particles.

These clouds then move, create precipitation, block the sun, and create many more effects. What is adiabatic cooling? Gasses are governed by the ideal gas formula, which is $PV = nRT$, with P being pressure, V being volume, n is the number of moles, R being a constant, and T being temperature. By changing one variable, and holding 2 constants, the fourth one will inevitably change in response to keep the equation balanced. So, adiabatic cooling occurs when the pressure and volume drops, which causes the temperature to drop too. This can be rationalized by assigning each particle in the gas a "heat value," and so by spreading the particles out, the heat per area of volume will drop.

There are many types of clouds, all depending on the conditions of the air that the clouds formed in. Some of the most notable are Cumulonimbus clouds, which are thunderclouds, generated by masses of cool descending air and warm ascending air meeting. Along with the Nimbostratus (and a few other rare exceptions), these are the only cloud types that generate precipitation. Clouds can also form at varying heights, although all occurring in the troposphere (with a few exceptions).

There are also many types of precipitation, all dependent on the temperature of the air masses that the precipitation is falling through.

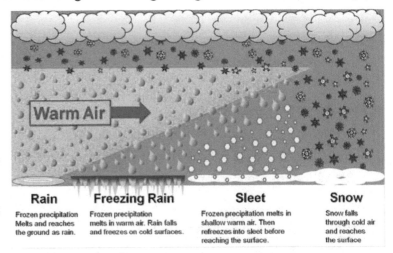

Figure 4.3-1 Rain and snow formation. Source: National Weather Service

4.4 Earth Layers Starburst by Jessica Fan

Materials:

1. Starbursts (color doesn't matter as much, just make sure you have the correct number)
   a. 1 yellow
   b. 1 pink
   c. 4 red
   d. 2 dark red
2. Tray
3. Gloves

Procedures:

1. Using each of your Starbursts to represent a layer of the earth, make a model of the layers of the earth. You can make it any way you want, but below is a sample model you can build.
2. Roll one Starburst into a ball. This represents the inner core.
3. Take a Starburst and flatten it out, then wrap it around the ball you just made. This represents the outer core.
4. Take three Starbursts and flatten them out. Wrap it around the ball you just made (this layer should be significantly thicker than the other layers. If not, add a few more layers).
5. Finally, make a thin sheet using the two Starbursts you have left, and wrap it around the ball. This represents the crust, which is the thinnest of all the layers.

Safety:

Be sure to wash your hands after touching the Starbursts.

Don't eat the Starbursts after molding them because they might get dirty.

Experiment Questions

1. Why do you think there were different numbers of Starburst for each layer of the earth?
2. Do you think the colors meant anything?
3. What layer turned out to be the thickest? (The Mantle was supposed to be the thickest)
4. What are some other ways of modeling the earth layers without forming the Starbursts into a ball? Perhaps you could stack them?

5. What differences do these models show?

Background and Applications:

Have you ever wondered how big the earth truly is? Has a scientist dug down deep into the core of the earth before? Unfortunately, the earth is huge, and the deepest that geologists have ever drilled down to is its topmost layer: the crust.

There are various layers in the earth, which can be differentiated through their properties. There are layers based on mechanical properties such as viscosity: lithosphere, asthenosphere, lower mantle (also known as mesospheric mantle), outer core, and inner core, as well as layers based on chemical composition: Crust, Mantle, Outer Core, and Inner core. Today, we will be discovering the characteristics of the layers based on chemical composition.

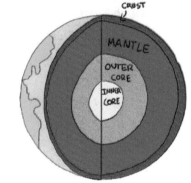

Figure 4.4-1 Earth layers

Our journey begins on the surface of the earth, with the crust. This is the thinnest layer of the Earth (5km-100km) and consists primarily of the elements, oxygen, silicon, and aluminum. There are two types of crust: continental and oceanic. The continental crust is thicker and lighter, while the oceanic crust is thinner and denser.

We then travel deeper into the earth, to the second layer. The mantle is the thickest layer of the Earth (2,900 km thick). It's made of molten rock, and has plasticity, allowing it to flow underneath the crust. This causes convection currents, driving tectonic movement, continental drift, and seafloor spreading. The mantle is equivalent to 84% of Earth's total volume and has temperatures ranging from 200 degrees to 4000 degrees Celsius. The mantle is made of two parts: the upper mantle and the lower mantle.

The outer core is composed of iron and nickel. Because the temperatures are so high, the outer core is liquid. It's 2,200 kilometers thick and 4,500 degrees Celsius to 6,000 degrees Celsius. The outer core is in charge of the earth's magnetic field, which protects us from cosmic radiation.

Finally, we travel to the deepest layer: the inner core. It is a solid mass of iron that is 5,200 degrees Celsius. Despite the high temperatures, the inner core is not liquid. In fact, the inner core is solid because of the immense pressure (3.6 million atmospheres (atm)) from all the layers above it.

After exploring the layers of the earth, we now understand the composition of the wonderful world we live in.

STEMJUMP   **Supervision by an adult is recommended.**

4.5 Greenhouse Gases by Eric Zhang

Materials:

1. 2 aluminum foil trays
2. Plastic sheet wrap
3. 200 mL water
4. Thermometer

Procedures:

1. Bring the aluminum foil trays outside.
2. Pour in 100 mL of water in each tray.
3. Using the thermometer, measure the starting temperature in the aluminum foil trays.
4. Use plastic sheet wrap to wrap the top of one tray.
5. After 2 hours, poke a small hole in the plastic sheet wrap and check the temperature inside the aluminum foil tray by sticking a thermometer into the small hole. Then, check the temperature inside the aluminum foil tray without the plastic sheet wrap.
6. Calculate the difference in temperature between the two trays.

Safety:

Remember to put sunscreen on when working outside.

Experiment Questions:

1. Why is the temperature of the trays different? What effects may the plastic wrap have?
2. What would happen if we used a large piece of paper instead of the plastic wrap? Would the temperature be different?
3. How does this model relate to earth's greenhouse effect?
4. Why are greenhouse gases important for life on earth?
5. Why did the Industrial Revolution coincide with increasing greenhouse gas concentrations?

Background and Applications:

Demonstrate the greenhouse gas effect so that students can visualize it and apply it to the earth's greenhouse gas effect.

127

Earth's atmosphere is composed of multiple layers of air that surround the earth. From lowest altitude to highest altitude, these layers are: the troposphere, the stratosphere, the mesosphere, the thermosphere, and the exosphere. Greenhouse gases trap heat, and it surprisingly slightly cools the stratosphere. This effect is known as the greenhouse effect.

A good analogy for earth's greenhouse effect is a greenhouse structure for growing plants. Greenhouses are used commercially to create an optimal temperature, humidity, and water level for various plants to grow. The building maintains a warmer temperature relative to the outside temperature by utilizing a transparent roof, which allows radiation from the sun to pass through and warm up the ground and air inside the greenhouse. The roof and walls effectively "trap" this heat inside because the heat cannot pass through as easily as the incoming solar radiation.

Earth's greenhouse gasses perform the same function as the special roof of greenhouse structures: trapping incoming solar radiation to hold a warmer temperature within the atmosphere. This allows life to flourish in these warmer temperatures; if it was not for greenhouse gases, we would face much more extreme conditions. Take our own moon, for example. It does not have an atmosphere or greenhouse gasses, and its daytime temperatures can reach 121 degrees Celsius (250 °F) while its nighttime temperatures can drop to -133 degrees Celsius (-208 °F). Conversely, Venus's atmosphere is mostly carbon dioxide, a greenhouse gas. Scientists believe that Venus is uninhabitable for this reason; the concentration of greenhouse gases in its atmosphere, coupled with its proximity to the sun, create temperatures that average around 465 degrees Celsius (870 °F). Earth's concentration of various greenhouse gases creates a window in which life can flourish, and it is critical to maintain this balance.

Let us follow a full day-night cycle to divulge the mechanisms behind the greenhouse effect. The concept is very similar to how greenhouse structures for plant cultivation work.

During the daytime, the sun's radiation travels through space to reach earth. The sun's radiation typically comes in the form of **ultraviolet, visible,** and **infrared** radiation. Since the sun is around 150,000,000 kilometers (93,000,000 miles) from earth, and the speed of light is around 300,000,000 meters per second, we find that it takes around 8 minutes for sunlight to reach earth.

Solar radiation can be **absorbed** and **scattered** in the atmosphere. It can be **reflected** by clouds back up towards space, and some of it is **diffused** through the cloud towards the ground. Once the radiation hits the ground, it can be reflected upwards or absorbed by the ground. As the day goes on, the radiation warms the earth's surface.

At night, there is no incoming solar radiation. **Longwave** or **infrared** radiation is released from the surface and rises up, but much of it is trapped within the atmosphere due to the greenhouse gases in our atmosphere. This is why temperatures at night usually do not dip as low as temperatures like our moon's during the night.

Figure 4.5-1 The greenhouse effect: the arrows represent how radiation is trapped by greenhouse gases in the atmosphere. Source: National Aeronautics and Space Administration

We will now run through the major greenhouse gases, their sources, and their comparable effects. In decreasing abundance, as defined by how many molecules of each gas exist in the atmosphere, they are: water vapor (H_2O), carbon dioxide (CO_2), methane (CH_4), nitrous oxide (N_2O), ozone (O_3), chlorofluorocarbons, hydrofluorocarbons, perfluorocarbons, sulfur hexafluoride (SF_6), and Nitrogen Trifluoride (NF_3).

Water vapor primarily comes from **evaporation** from bodies of water, as well as transpiration from **plants**.

Carbon dioxide is an **anthropogenic** greenhouse gas, meaning that its emission can be largely attributed to human activity. More specifically, since the **Industrial Revolution** in the 18th century, carbon dioxide levels, as well as other greenhouse gases emitted by industrial activity, have risen significantly. Carbon dioxide is mainly produced from fossil fuels, transportation, and industry.

Methane is the second most common anthropogenic greenhouse gas. Although it is less abundant than carbon dioxide, it is 28 times more potent as a greenhouse gas. However, it also has a shorter lifespan than carbon dioxide. Methane is mainly produced from fossil fuels, agricultural activity, and landfills.

Since 1900, the global average temperature of earth has risen by around 1 degree Celsius (2 degrees Fahrenheit). Many climatologists believe that the rise in greenhouse gas emissions in the last few centuries contributes to **global warming**, the long-term heating of planet earth. **Climate change** is a broader term that describes the change in earth's weather and temperature patterns over time.

To combat this, some governments and organizations have started some initiatives to reduce greenhouse gas emissions. **Clean energy** comes from **renewable** sources and does not emit greenhouse gases; it has become used more commonly in many countries. Wind power, solar power, hydroelectric power, and nuclear power are considered sources of clean energy, and they provide an alternative to **non-renewable**, greenhouse gas-emitting fossil fuels.

There have been some international efforts to reduce greenhouse gas emissions and combat its effects on global warming. The **Kyoto Protocol** was signed in December 1997, and it was eventually ratified by 147 states. Its goal was to set targets for individual countries to each reduce their greenhouse gas emissions, so that the greenhouse gas concentrations in the atmosphere could be stabilized.

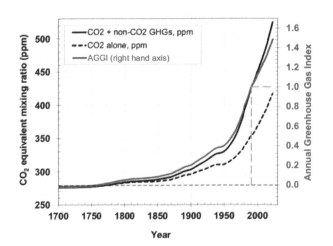

Figure 4.5-2 Greenhouse gas emissions have been increasing over time. Source: NOAA Global Monitoring Laboratory

The **Paris Agreement** is an international agreement signed in 2016 to mitigate the effects of greenhouse gases on climate change. Its goal is to limit the rise of global temperatures to under 2 degrees Celsius (3.6 °F) more than pre-industrial concentrations. It requires each participating country to record and submit a **Nationally Determined Contribution** every five years to detail their efforts towards reducing greenhouse gas emissions. It has essentially replaced the Kyoto Protocol as the main international policy on climate change.

Greenhouse gases are crucial to life on earth as we know it due to their unique effects of trapping heat within the atmosphere. However, the rise of anthropogenic greenhouse gas emissions has contributed to increased temperatures over the last few centuries, and many groups of people have begun to form solutions and alternatives to greenhouse gas-emitting human activities. As we move forward, we hope that this section has given a clear understanding of greenhouse gases and their effect.

Supervision by an adult is recommended.

4.6 The Magnificent Moon by Eric Zhang

Materials:

1. 1 white ping pong ball
2. 1 of any rod-like objects, for example:
   - A pencil
   - A ruler
   - A stick
3. Tape
4. Flashlight

Procedures:

1. At the tip of your rod, tape the white ping pong ball on.
2. Turn off the lights.
3. Turn on the flashlight. Point it at yourself and place the flashlight around 10 feet away from you.
4. Hold the rod horizontally at arm's length with the ping pong ball facing away from you.
5. Rotate in place. Observe how the flashlight's light illuminates the ping pong ball.
6. (Optional) draw diagrams of the ping pong ball as you rotate.

Safety:

Do not touch the flashlight for too long because it gets hot.

Do not poke yourself or others with the rod.

Experiment Questions

1. If the moon revolves around the earth twice as fast, would there be full moons more often? Why or why not?
2. With what you have learned about moon phases and the moon's revolution around earth, how do you think lunar eclipses happen?
3. Leslie's boss gives paychecks based on the phases of the moon. It was a waxing gibbous moon yesterday, December 7. If her paycheck arrives every new moon, around which date will she expect her paycheck to next arrive?

Background and Applications:

Model the phases of the moon as well as its revolution around earth.

We hope that this experiment gives a good model of moon phases. In the section, we will explore the moon's composition, phases, cycle, eclipses, and many other of its aspects. Located more than 230,000 miles away, the moon is almost always visible from earth on clear nights, and it has a profound impact on our daily lives, animals, and natural processes on earth.

There are many theories on how the moon was formed, but the **Giant-impact hypothesis** is one of the most widely accepted theories. It states that the moon was formed around 4.5 billion years ago when a smaller planet or object smashed into the earth. This created a field of debris orbiting earth, which eventually coalesced into the moon. moon rocks sampled from NASA's Apollo program have been studied to have similar isotope ratios to rocks on earth, supporting this hypothesis.

A **satellite** is an object that orbits a larger object. Many **artificial satellites** orbit earth and provide us with GPS, radio signaling, weather forecasting, and other technological services. However, the earth has only one **natural satellite**: the moon. The moon takes 27.3 days to revolve around earth; however, it takes 29.5 days from a new moon to the next new moon. Why is this? Well, as this period of time passes, Earth also **revolves** around the sun, so the moon must revolve a smidge more to "sync" with the earth's new position respective to the sun so that they are aligned to form a new moon. We will explore how these revolutions control the phases of the moon shortly. Feel free to preview the phases of the moon shown below in Figure 4.6-1.

Figure 4.6-1 The phases of the moon. Source: NASA

Since the moon is illuminated by the sun, it is always illuminated on one side of itself. As the moon revolves around the earth through different positions, our angle of viewing towards the moon also changes, and we see that different fractions of the moon are illuminated; these are the different phases of the moon.

The phases of the moon are: New, Waxing Crescent, First Quarter, Waxing Gibbous, Full, Waning Gibbous, Last Quarter, and finally Waning Crescent. When the moon is waxing, that means the portion illuminated by the sun begins to grow from right to left. When the moon is waning, the portion not illuminated by the sun begins to grow from right to left; that is, the darkness increases from right to left.

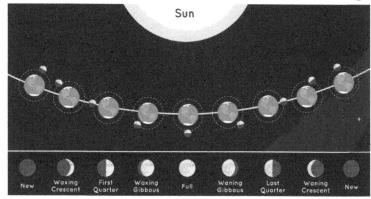

Figure 4.6-2 The moon in revolution. Source: NASA

Recall that it takes around 29.5 days from one new moon to the next new moon. For simpler calculations, let us round this to 28 days, or 4 weeks. This means that the first quarter phase (when the right half of the moon is illuminated when viewed from earth) occurs around 28 divided by 4 = 7 days, or 1 week after the new moon. The time between a full moon and a new moon is around 28 divided by 2 = 14 days, or 2 weeks. The last quarter phase, as the name suggests, occurs around 3 weeks after a new moon or 1 week before a full moon.

The moon causes tides on earth. A tide is the oscillation of the sea water level with a period of 12 hours. The high tide is the highest water level of the ocean in this cycle, and the low tide is the lowest water level of the ocean in this cycle. 12 hours elapse between two high tides, 12 hours elapse between two low tides, and 6 hours elapse between a high tide and a low tide. The tides are caused by the moon's gravitational pull on the closest and the farthest sides of the earth to the moon. The oceans essentially "bulge" on opposite sides. The earth and the moon are tidally locked, meaning that the same side of the moon always faces earth.

When the sun, the moon, and the earth are aligned during a spring tide, the sun's gravitational pull adds on to the moon's gravitational pull, making the highest high tides and the lowest low tides. Spring tides occur during full and new moons. In contrast, when the moon's gravitational pull is perpendicular to the sun's during a neap tide, the difference in water level between the high and low tides is the least. Neap tides occur during the first and third quarter moons.

Now, let's learn about how the moon can be observed. Even though it is visible to the naked eye on a clear night, some people prefer to use binoculars or telescopes that magnify the moon so that they are able to see more details of the moon. NASA's Lunar Reconnaissance Orbiter orbits the Moon and takes close up, detailed images of the moon.

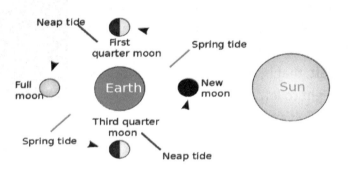

Figure 4.6-3 Tides. Source: commons.Wikimedia.org

There are 3 main landforms that exist on the moon's surface: maria, highlands, and craters. Maria are the large dark patches on the moon that were formed by lava that flowed into basins. They are plains composed of basalt. Highlands are the lighter parts of the moon that surround maria and contain many craters. These craters were formed when asteroids and comets impacted the surface of the moon.

The moon is not only valuable in the study of our planet and solar system, it is a magnificent spectacle of the night sky that has cultural significance to many people. We hope that his demonstration and textbook section have provided quality insight to earth's natural satellite.

4.7 Stunning Salinity by Eric Zhang

Materials:

1. 500 mL water
2. 50 g salt
3. 300 g ice
4. 1 bucket
5. Hot plate (optional)
6. Scale (optional)

Procedures:

1. Fill up the bucket with 500 mL of water.
2. Add in the 50g of salt.
3. Stir the solution until it dissolves.
4. Now, add in the 300g ice.
5. Let the ice melt.
6. Measure the initial mass of the saltwater (optional).
7. Heat the saltwater over a hotplate (optional).
8. Measure the final mass after all of the water has dissolved (optional).

Safety:

Do not touch or taste the water solution. Wear goggles throughout the experiment.

Experiment Questions

1. Calculate the salinity of the water solution before and after the ice melted. What do you notice?
2. If sea ice continues to melt, what effects will it have on ocean salinity? How will this impact marine organisms and ecosystems? What are some ways species may adapt?
3. How are newer irrigation methods combating the issue of soil salinity?
4. Researchers gather 800 grams of lake water and find that it has 24 grams of salt dissolved in it. Calculate the salinity in parts per thousand (ppt). What does this say about the lake water?

Background and Applications:

Demonstrate one of the effects of melting sea ice on the environment.

We hope that this experiment gave or will give a nice demonstration on how salinity can be altered, and the potential effects of raising or lowering salinity. Salinity is the amount of salt that is dissolved in a body of water, which forms the basis of ecosystems and the physical properties of a body of water. We will explore the properties of salinity, factors affecting salinity, the impact on organisms, as well as the human impact on salinity's role in our planet. We will begin with a case study on The Dead Sea, a salty lake located on the border between Israel and Jordan.

Scientists studying salinity need a way to measure it before conducting studies and experiments. They most often use the unit parts per thousand (ppt), denoting the mass of dissolved salts (in grams) in 1000 grams (1 kg) of saltwater. Similarly, parts per million (ppm) denotes the mass of dissolved salts (in grams) in 1000000 grams (1000 kg) of saltwater. In other words:

$$x\ ppt\ =\ \frac{x\ grams\ of\ salt}{1000\ grams\ of\ saltwater}$$

$$x\ ppm\ =\ \frac{x\ grams\ of\ salt}{1000000\ grams\ of\ saltwater}$$

A Case Study: **The Dead Sea**

Due to its extreme salt content of around 34%, very little life exists in the waters of the dead sea. However, during periods of time with significant precipitation, the salt content can drop to less than 30%. *Dunaliella*, a species of algae and halobacteria have been discovered during one of these periods of time in the early 1980s. Additionally, biofilms of prokaryotic bacteria and archaea have been discovered near the floor of the Dead Sea. This body of water serves as a reminder of the narrow windows in which life may exist.

The Dead Sea is now a popular tourist attraction. Some individuals believe that the unique saline mud and water of the Dead Sea carry health benefits for humans. Scientific research on human health is also being done at this location.

Figure 4.7-1 The Dead Sea viewed from space. Source: NASA

We can also use this formula to easily convert percent composition of dissolved salts by mass to parts per thousand and parts per million:

$$x \; ppt \; = \; 0.1x \; \% \; salt = 1000x \; ppm$$

Oceans have a salinity of around 35 parts per thousand. The United States Geological Survey (USGS) classifies different bodies of water by its salinity, as detailed by the following table:

Salinity in ppt	USGS Classification
<1 ppt	Freshwater
1 - 3 ppt	Slightly Saline Water
3 - 10 ppt	Moderately Saline Water
10 - 35 ppt	Highly Saline Water

Of course, there are other systems of classification that different organizations or governing bodies may utilize.

Salinity is not merely a number; people must have instruments and methods to measure it. A simple way to do this at home is to heat a sample of saltwater until the water dissolves and only salt is left behind. Take the remaining mass of salt and divide by the original mass of the liquid to find the percentage of dissolved salt, and convert to ppt or ppm if desired. During your experiment, you may try using a hotplate to isolate the salt and calculate the salinity.

A salinometer is any device that measures the salinity of a solution. Two interesting and common instruments include the hydrometer and the electrical conductivity meter. A hydrometer uses the density of a liquid by measuring the force of buoyancy. It calculates salinity from the temperature, salinity, and water density curve. An electrical conductivity meter passes an electric current into saltwater, and calculates the salinity based on the principle that conductivity increases with salinity.

Figure 4.7-2 A Salinometer (Electrical Conductivity Meter). Source: Photograph by Mike Peel (www.mikepeel.net), CC BY-SA 4.0

Salinity can be affected by location, precipitation, and human activities. A salinity gradient is present at estuaries, where a freshwater stream or river meets the ocean. As water depth increases, salinity increases too. These unique conditions of salinity contribute to the valuable biodiversity that exists in estuaries.

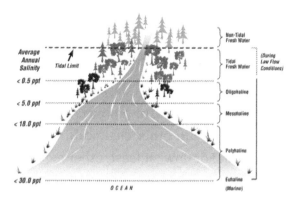

Figure 4.7-3 Salinity gradient in an estuary. Source: The United States Environmental Protection Agency (EPA)

Additionally, sea ice affects ocean salinity. Recall that water freezes and leaves salt behind. Therefore, the formation of sea ice during the colder months will slightly increase salinity in the nearby waters by removing freshwater, whereas the melting of sea ice during warmer ice will slightly decrease salinity in the nearby waters by mixing in freshwater. Some climatologists are concerned that with rising average temperatures, more sea ice will melt and decrease salinity of seawater near the poles. This may inhibit thermohaline circulation (deep-ocean currents that are driven by differences in the temperature and salinity of water) which could slow down the global conveyor belt and alter deep sea currents worldwide.

Organisms have adapted to the salinities of their environments. Saltwater environments provide a small problem for cells: they tend to make cells shrivel as water exits the cell by osmosis to equalize the salt concentration between inside and outside of the cell. Thus, in marine environments, animals have specialized organs to regulate the osmosis of water and the diffusion of ions in a process known as osmoregulation. For example, fish use their gills to remove excess salt. However, most marine organisms would not be able to survive in freshwater environments because their structures have been adapted to thrive in saltwater environments.

One interesting organism that spends different phases of its life in both freshwater and saltwater is salmon. Salmon lay their eggs in freshwater streams. As salmon develop, their body structures begin to allow them to live in saltwater. Most salmon will migrate to the ocean to live their adult life, before returning to their freshwater spawn to lay their eggs. This pattern of migration makes salmon an anadromous species.

Some governments around the world are looking at seawater as a potential source for drinking water and irrigation. However, removing the salt and minerals to make the water safe for consumption or irrigation requires the expensive process of desalination. The biggest desalination plants are located in the United Arab Emirates, Saudi Arabia, and Israel. It is not a coincidence that in these places, natural sources of freshwater, like rivers and springs, are uncommon.

There are a few challenges to desalination. Firstly, desalination is energy costly and therefore an expensive process to run. Additionally, the waste products of desalination contain brine (water with a very high concentration of salt) and other chemicals, which can build up and negatively impact the surrounding environment. For now, at least, other methods of acquiring freshwater for drinking and irrigation are cheaper and less environmentally harmful than desalination in most places.

Humans can impact soil salinity by using certain methods of irrigation, which can be detrimental to the agricultural productivity of a plot of land in the long term. One irrigation method known as flood irrigation is responsible for some of these cases. As the name suggests, in flood irrigation, more water than necessary is applied to the soil. When a large volume of water is evaporated, a large amount of salt is left behind. This increase in soil salinity can make the soils unable to grow crops after a few years.

Figure 4.7-4 Drip irrigation. Source: US Department of Agriculture

One way to prevent the loss of agriculturally viable land is to employ alternate methods of irrigation. Drip irrigation involves pipes that extend right above the soil, and it "drips" small amounts of water onto the soil. This conserves water uses while maintaining normal soil salinity for longer periods of time. While there are some disadvantages of drip irrigation, it is an innovation that has given us a more sustainable method of irrigation in agriculture.

It is certainly stunning that something as simple as dissolved salt in a volume of water can affect ecosystems, agriculture, and ocean currents around the world. We hope that his section provided an interesting and valuable perspective of our planet earth and its natural processes.

140

4.8 Exhilarating Earthquakes by Eric Zhang

Materials:
1. *2 approximately equally-sized books*
2. *1 set of Jenga blocks or similar blocks*

Procedures:
1. *Set the two books next to each other, laying flat.*
2. *Using the blocks, build a simple tower that lays on top of both of the books.*
3. *Now, begin to rub the books against each other, keeping them lying flat.*
4. *Vary the intensity of the shaking and observe how it affects the tower.*

Safety:
Wear close-toed shoes to protect your feet if blocks fall on the floor.

Experiment Questions:
1. Can you think of a way to make the tower stronger to withstand the force of the shaking books? What about its design?

2. In this experiment, what kind of tectonic plate boundary did the books represent? What kind of fault line was it?

3. If earthquake A had a Richter Scale magnitude of 4, and earthquake B had a Richter Scale magnitude of 7, how many times more intense (ie. amplitude of seismic wave) is earthquake B compared to earthquake A?

4. Research Question: What are some examples of convergent boundaries in the world?

Background and Applications:
Gives a demonstration of fault lines, transform plate boundaries, and the impact of earthquakes on infrastructure.

This experiment was meant to model how an earthquake happens and its effect on buildings and cities. Earthquakes can range from barely noticeable tremors in the ground to severely damaging shaking. In this section, we will explore the mechanisms of earthquakes, their measurement, and their real-world impacts.

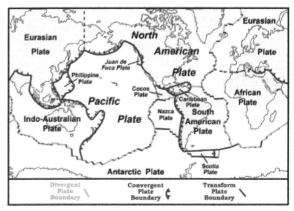

The earth is composed of many layers, including the **inner core**, **outer core**, **mantle**, and **crust**. We will focus on the crust because it is the layer on which forces act to cause earthquakes.

Figure 4.8-1 Earth's tectonic plates. Source: United States National Park Service

The crust is the outermost layer of the earth, and it is split into many **tectonic plates**. These are essentially large pieces of the crust that move slowly. They sometimes interact with each other, which is where some of the trouble comes from. If two plates are moving away from each other, the boundary is known as a **divergent boundary**. If two plates are moving toward each other, the boundary is known as a **convergent boundary**. If two plates are sliding against each other, the boundary is known as a **transform boundary**.

Fault lines can form when there is a break in the rock and they move against each other. All plate boundaries are faults, but not all faults are plate boundaries. When plates push, pull, or rub against each other, they store some **elastic potential energy** in the crust, which is released occasionally into the **kinetic energy** that drives earthquakes when the plates eventually slip. Divergent boundaries cause relatively minor earthquakes in faults known as **normal faults**. Most earthquakes can be attributed to the colliding of plates in convergent boundaries in faults known as **reverse faults**. Transform boundaries cause some earthquakes in faults known as **strike-slip faults**.

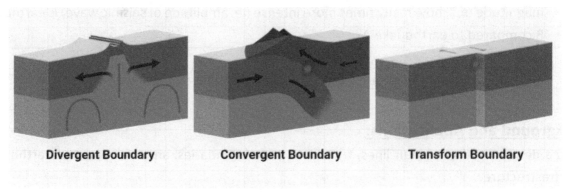

Figure 4.8-2 Types of faults. Source: Domdomegg, CC BY 4.0

One very well-known example of a fault is the **San Andreas Fault**, which runs through California between the Pacific and North American Plates. Since these two plates slide parallel to each other in opposite directions, it is a transform boundary, and the fault is a strike-slip fault. This fault is responsible for many earthquakes in this region, many of which were severe and destructive.

The waves of **acoustic energy**, or **seismic waves**, originate at the **hypocenter** (also known as the **focus**) of the earthquake. The **epicenter** is the location on the surface of the earth directly above the hypocenter. There are many types of **seismic waves**, including **p-waves**, **s-waves**, and **surface waves**. P-waves, or primary waves, are the fastest waves. They can move through solids, liquids, and gases. S-waves, or secondary waves, can only move through solids. Surface waves only travel on the surface of the earth (as the name suggests), and they travel more slowly than the **body waves** (P-waves and S-waves). The study of these seismic waves has provided researchers with very convincing evidence of the different layers of the earth and their properties. A **seismograph** (or **seismometer**) records the motion of the earth during earthquakes by drawing the seismic waves on a **seismogram**.

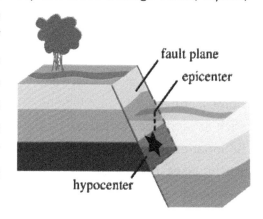

Figure 4.8-3 Earthquake diagram. Source: United States Geological Survey

Geologists use the **Richter Scale** to measure the intensity of earthquakes. This scale is **logarithmic**, meaning that a 1-point increase on the Richter Scale corresponds to a 10-fold increase in earthquake intensity (as defined by the amplitude of the seismic wave). Here is a formula that can model this relationship:

$$M = log(\frac{I}{I_0}) \qquad\qquad \text{or } I = I_0 \times 10^M$$

where M = Richter Scale magnitude, I = intensity of the earthquake being measured, and I0 = a constant representing the intensity of an earthquake with Richter Scale magnitude of 0. If an earthquake has a Richter Scale magnitude that is x higher than another earthquake, it is 10x more intense.

Sometimes, during foreshocks, smaller earthquakes can precede a larger earthquake event known as the mainshock. Similarly, aftershocks may come after the larger earthquake event. Foreshocks and aftershocks usually occur in the same area as the mainshock.

Earthquakes can be deadly. They can destroy buildings and highways, displace people from their homes, or even trap people beneath rubble. They can also trigger other dangerous natural events, such as landslides or tsunamis. The area's most prone are located near faults and tectonic plate boundaries, and they may have older infrastructure that isn't strong enough to withstand earthquakes. Around the world, wealthier cities near fault lines have been able to begin designing new buildings with the threat of earthquakes in mind.

We hope that this section provided an exhilarating tour of the study of earthquakes, its causes, and its effects. As we move forward, the new generation of scientists, geologists, and engineers across the globe will be faced with the important task of combatting the destruction caused by earthquakes.

4.9 Epic Erosion by Eric Zhang

Materials:

1. 1 tub
2. Sand to fill up the tub halfway
3. A fan (can be electric or just handmade)
4. 500 mL water

Procedures:

1. Fill up the tub to half of its height with sand.
2. Turn on the fan, or wave it with your hands if it's not electric.
3. Point the fan to the tub.
4. Observe how the wind affects the erosion of sand.
5. Now, turn off the fan, or stop waving it with your hands.
6. Begin to pour water into the tub, concentrated in 1 area.
7. Observe how water affects the erosion of sand.
8. Put away materials and clean up the work environment
9. Complete the post-experiment open-ended questions.

Safety:

Wear goggles to prevent sand from getting into your eyes.
Be careful with the sand, it is heavy.

Experiment Questions

1. Think about the sand dunes of the Sahara Desert. How do you think wind erosion may have formed these? How will wind erosion continue to affect these?
2. After watching the demonstrations of erosion, give a possible reason why the Grand Canyon formed. Also, list out possible forms of erosion that may still affect the Grand Canyon today.
3. What do you think would happen if there were tree roots in the sand in our experiment? How would that affect the amount of erosion that occurs? What does this reveal about the long-term effects of deforestation around the world?

4. Think about how the factors that play into erosion and the effects it may have on habitats and ecosystems. Write a persuasive letter to your representative on the ways to prevent and combat the effects of erosion.

Background and Applications:

Create a visual demonstration of the effect of water and wind erosion for a conceptual understanding of the fundamental processes that shape our planet.

After you have completed this brief demonstration of the causes and effects of erosion, let us dive into the intriguing processes behind erosion. **Erosion** is the transport of pieces of rock, soil, or sand from one place to another. **Water** and **wind** are two common causes of this movement that we explored in the experiment. Some other causes of erosion include **gravity** and **glaciers**. In this section, we will take a tour around the world to contextualize these forces of erosion in action.

Water erosion is the transport of soil or rock particles by rivers, rainfall, and various sources of **runoff**. It is often aided by gravity; water generally flows from higher regions to lower regions, tracing out the particular areas of erosion that the water acts on.

One of the classic examples of water erosion is the **Grand Canyon** in the state of Arizona. Over millions of years, the Colorado River eroded the Colorado Plateau. As a result, visitors are exposed to the layers of ancient rock; water erosion has provided a beautiful glimpse of our planet's history.

Figure 4.9-1 The Grand Canyon and the Colorado river. Source: US National Park Service

In addition to rivers and streams, water erosion can be facilitated by runoff and rainfall. Of course, this would not be possible without the effects of gravity, causing rain to hit soil and rock at high velocities and slowly erode pieces of it away. Unfortunately, modern agricultural practices have increased the susceptibility of land to **soil erosion**. This reduces the ability to grow vegetation, degrading the ecosystems in the area. Eventually, soil erosion can lead to **desertification**, the loss of vital minerals and nutrients in topsoil required for vegetation to grow. There are four main types of soil erosion caused by runoff and rainfall:

Splash erosion occurs when raindrops hit the soil, and soil particles are broken up and flung into the air. In fact, these particles can be displaced up to 0.6 meters into the air!

Sheet erosion occurs when a mass of water removes soil particles, especially from the nutrient-rich topsoil, as it moves along the surface. The roots of trees and other plants help tremendously to prevent this form of soil erosion.

Rill erosion occurs when water enters the crevices or the depressions of soil, forming a small stream, or rill. As water moves through these rills, it erodes some of the soil particles away from its original position.

Gully erosion occurs when smaller water channels cut a larger channel deeper than 30 centimeters, and soil particles are eroded through these deep channels of moving water.

Wind can also be a powerful source of erosion. The sand dunes of the Sahara Desert were shaped by winds, primarily the **Harmattan** winds. These winds carry fine particles of sand along and form the massive sand dunes that characterize the region.

However, wind erosion has proven to be dangerous at many points in history. During the 1930s in America, the soil degradation from agriculture and severe drought brought about several severe dust storms in a period known as the Dust Bowl. Winds eroded and carried these particles through the air, destroying the air quality and preventing many families from running their farm operations. This event serves as a grim reminder of the importance of soil health and the capacity of wind erosion.

Figure 4.9-2 The dust bowl. Source: United States Department of Agriculture

Let us make our way into Alaska, where glacial erosion shapes the mountains and valleys in the region. The bottom portion of a glacier contains many rocks and sediments. As glaciers slowly inch forward, this rough bottom grinds against the bedrock and erodes away more sediments. This process is known as **abrasion**. When glacial ice melts, the rocks and sediments trapped in it are released and **deposited** into waterways and other places, where they will continue their journey. The sediment deposited by a glacier is known as **glacial till**, and it can clump together to form **moraines**. As we close this section, think about how erosion has become so ubiquitous in the formation of landforms and the occurrences of natural events. Whether by glaciers, rivers, runoff, or wind, erosion is a powerful force that has shaped our planet.

4.10 Solar Oven by Brian Lu

Materials:

1. Cardboard pizza box
2. Scissors
3. Aluminum foil
4. 1 roll of clear tape
5. Plastic wrap
6. Black construction paper
7. Newspaper
8. Something to hold up the lid (ex: skewer…)
9. Thermometer

Procedures:

1. Pop up the flap and tightly wrap foil around the flap, and then tape it to the back.
2. Tape two layers of plastic wrap to the hole in the top of the box. Make sure it is airtight.
3. Line the bottom of the box with black construction paper.
4. Add foil around all edges on the inside of the box and tape - the box should still be able to close.
5. Take the solar oven outside and adjust the flap so the most sunlight reflects off it.
6. Use something to prop up the solar oven.
7. Make a s'more. Graham cracker, marshmallow, chocolate, graham cracker.
8. Wait!

Safety:

This may get very hot. Be careful when removing food from the oven and when heating things using solar power.

Experiment Questions

1. What is the principle by which all electric generators function?
2. Why is nuclear energy considered green?
3. Where all energy sources on earth originate from?

Background and Applications:

Unless you've been living under a rock, you've heard of the climate crisis and how we need to switch to "green" energy. This switch is already in progress, with more and more renewable energy sources coming online every year. In this experiment, we will discuss the principles behind wind, solar, and many other energy sources.

First, let's talk about how energy generation works. Energy cannot be created, only transferred from other forms. In almost all of our energy generation types, we are taking stored potential energy (chemical/nuclear) and transferring it into electrical energy. All potential energy is stored in the form of mass, which is dictated by the equation $E=MC^2$. How this works is that the bonds between atoms, and subatomic particles have mass, and that mass is translated into energy.

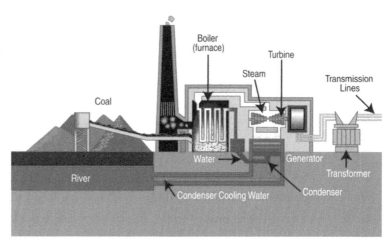

Figure 4.10-1 Electrical generation. Source: United States Tennessee Valley Authority

So, before we get into specifics of how solar energy specifically works, let's talk about how non-renewables, like coal, oil, and natural gas work (nuclear is an exception). All electrical generation sources work by spinning a dynamo, with one coil of copper wire being spun inside the other. Because electric wires generate electric fields, and can be controlled by current, moving the said electro-magnetic field will generate an electrical current. The majority of power types will use fuel to boil water, which then is passed through a turbine, which spins the dynamo. The only exception to the steam turbine is diesel generators, which use a series of pistons connected to a drive shaft to spin dynamos.

Now, let's talk about renewable energies. Before we start with the "traditional" renewables, let's talk about nuclear-fission energy. Nuclear-fission energy is technically non-renewable, but is generally considered "green" because it generates minimal amounts of CO_2. Nuclear plants also heat water to spin turbines, but through the fission of unstable materials, generally Uranium-235.

Fission works by using neutrons to trigger chain reactions of fission. First, a neutron is introduced (generally through decay), which strikes another atom, which releases more neutrons, and so on and so forth.

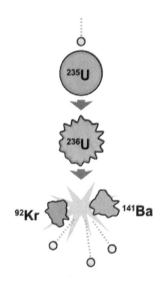

Figure 4.10-2 Nuclear-fission. Source: Wikimedia Commons

There is also the concept of Nuclear-Fusion energy, which is currently under development, which works by fusing atoms together (generally deuterium and tritium, which are isotopes of hydrogen). Fusion is very efficient and clean, if it can function at or close to its theoretical potential, but it also requires insane amounts of energy put in to get that energy output. Currently, fusion requires more energy input than it can produce.

Now, let's talk about wind and solar. The wind is pretty simple, using the natural air currents to spin a wind turbine, which spins a dynamo. Solar is a bit more complicated. First, there are solar panels, which are definitely out of the scope we can talk about here, but just know they take incoming electromagnetic radiation and transfer it directly into electricity. Then, there's solar-thermal, where solar radiation is used to heat water or some other thing to generate energy, which we will be testing today.

4.11 Outstanding Ozone by Eric Zhang

Materials:

1. 1 clear tub
2. 1 sheet of bubble wrap (may substitute with tissue paper if necessary)
3. 1 flashlight
4. 1 pair of scissors

Procedures:

1. Lay down the tub.
2. Cover the top of the tub with bubble wrap.
3. Shine the flashlight towards the top of the tub. Observe how the light is illuminated inside the tub, underneath the bubble wrap.
4. Cut a hole in the bubble wrap with a diameter of around 2 inches.
5. Shine the flashlight towards the top of the tub. Observe how the light is illuminated inside the tub, underneath the bubble wrap, and underneath the hole.

Safety:

1. Do not run with scissors in your hand.
2. Be careful when using scissors to cut the hole in the bubble wrap. Make slow movements to avoid unexpected injury.
3. Do not shine the flashlight in eyes to avoid eye injury.

Experiment Questions:

1. How does the bubble wrap affect how light passes through it?
2. What do you think would happen if we removed bubble wrap altogether?
3. The cut hole in the bubble wrap represents the Antarctic ozone hole. In this area of the earth, are levels of UV radiation greater or less than the average levels of UV radiation? Why or why not?
4. What types of atoms and molecules deplete ozone?
5. What did the Montreal Protocol accomplish?

Background and Applications:

Gives a simple model of how light can be scattered, representing the ozone layer's effect on UV radiation as well as the detrimental impact of an ozone hole.

Ozone is one of the most important gases that protect life on earth. Its outstanding properties that shield UV radiation from us are undoubtedly crucial to preserve.

Ozone's chemical formula is O3, making it a triatomic particle. Its structure is trigonal planar, and its Lewis structure representation is the following:

Figure 4.11-1: Ozone Lewis structure. Source: Ben Mills, Public Domain

Ozone forms when a ray of UV light strikes oxygen gas (O_2) into O and O individually. Each O atom collides with molecular oxygen gas, creating O_3. A rudimentary equation for the formation of ozone is:

$$O + O_2 \rightarrow O_3$$

Let's explore what UV radiation is and why it is crucial to block it. UV radiation, also known as **ultraviolet radiation**, is a type of radiation on the **electromagnetic spectrum** that has a longer wavelength than **x-ray radiation** but a shorter wavelength than **visible light**. Here is the electromagnetic spectrum, for reference:

THE ELECTROMAGNETIC SPECTRUM

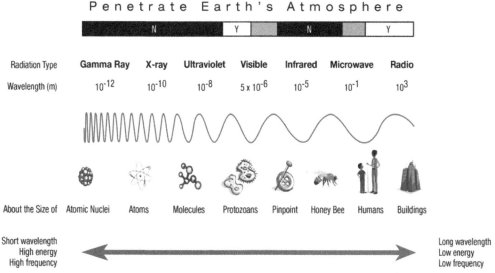

Figure 4.11-2 Electromagnetic spectrum. Source: National Aeronautics and Space Administration

Because UV has a higher **frequency** and shorter **wavelength** than visible light, you can think of it as being a "sharper" light than the visible light that we are able to see. UV light cannot be seen with the human eye, but it can be seen with some specialized cameras. There are three main forms of UV radiation (listed from longest to shortest wavelength): UV-A, UV-B, UV-C.

As you may have heard, excessive UV radiation is harmful to one's skin. You may remember painful sunburns after a sunny day at the beach. Over time, these may sometimes develop into various forms of skin cancer if one does not wear sun protection but spends a lot of time in the sun. There is some research indicating that it may accelerate aging and lead to unhealthier skin. **Melanin** is a pigment in the skin that protects against UV radiation, but it doesn't completely account for the levels of UV radiation that is exposed to some people.

This is where ozone comes into play. Ozone is able to absorb UV rays and convert it into heat energy when it splits the ozone into a singular O atom as well as molecular O_2 gas. This is the reverse of the way ozone is formed; it is essentially a "cycle" so that the ozone will never be depleted under natural processes. We can model this breakdown with another rudimentary chemical equation:

$$O_3 + UV \rightarrow O_2 + O$$

How do scientists take measurements of the amount of ozone gas in the air? The **Dobson unit** is a convenient method of measuring ozone levels. It is defined as the total amount of ozone per unit area. More specifically, if you take a vertical column of the atmosphere, one Dobson would be equivalent to pure ozone with a thickness in the vertical column of 0.01mm. The average measurement of ozone is about 300 Dobson units.

However, **ozone depletion** threatens the existence of these important compounds that protect life on earth. One very notable example of this is the ozone hole that presides over Antarctica. The ozone hole drops below the ozone level of 220 Dobson units, and it dips to an average of about 100 Dobson Units.

This ozone depletion is caused by **chlorine** and **bromine** atoms interacting with ozone molecules in the **stratosphere**. Over time, this uses up the available ozone molecules and decreases its prevalence, causing depletion. Some of the larger sources of these ozone-depleting atoms include **chlorofluorocarbons** (CFCs), **hydrochlorofluorocarbons** (HCFCs), **methyl bromide**, and **halons**. Chlorofluorocarbons have been used for making various products such as refrigerants or aerosols, contributing to the depletion of ozone.

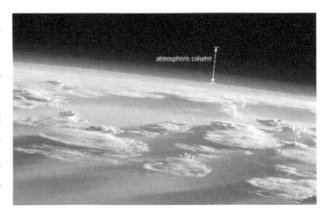

Figure 4.11-3 Vertical column of the atmosphere. Source: National Aeronautics and Space Administration

In 1987, the **Montreal Protocol** was signed and since then, 198 states and parties have ratified them, including all members of the United Nations. It restricted the usage of many ozone-depleting substances, including chlorofluorocarbons. It has proven to be quite successful—the use of ozone-depleting substances has gone down significantly worldwide, and the Antarctic ozone hole has begun to shrink.

The Montreal Protocol's success is a rare but important precedent in the reduction of environmentally harmful emissions. It serves as a reminder that given a challenge, humanity has the ability to set aside political differences and work together to resolve serious problems that threaten the well-being of humans and life on earth, such as ozone depletion. As you move on from this section, we hope you've gained a solid understanding of ozone's mechanisms of blocking UV light as well as its importance for life on earth. Knowledge underlies the power of the individual and the progression of human society.

4.12 Radial Velocity by Brian Lu

Materials:

1. *Small child*
2. *Whistle*

Procedures:

1. *Give your child the whistle and have them run in a straight line.*
2. *Place an observer in the middle of the line the runner will run.*
3. *Record what happens to the pitch of the whistle.*
4. *Repeat with different speeds (recommended to repeat with sprint, light jog, walk, and stopped).*

Safety

Don't trip and fall.

Experiment Questions

1. Why does the pitch of the sound change as the runner, well, runs?
2. How does this apply to finding exoplanets?
3. What is redshift?

Background and Applications

Ever wonder how astronomers detect planets that are millions of lightyears away? The majority of it is through the transit method, which we will cover later. In this experiment, we will be testing the radial velocity method of exoplanet detection. First, some background information on how stars function.

Stars are held in hydrostatic equilibrium by gravity and radiative pressure from nuclear fusion. The specifics of nuclear fusion are beyond the scope of this book. All you need to know is that hydrogen combines with itself in various ways to form helium (for the majority of a star's lifespan).

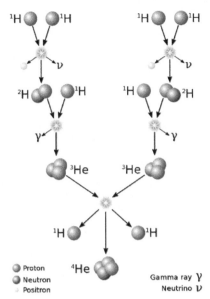

Figure 4.12-1 A diagram of the proton-proton chain. Source: Sarang - Own work, Public Domain,
https://commons.wikimedia.org/w/index.php?curid=51118538

This generates a lot of energy. A LOT of energy, about 0.7% compared to theoretical predictions by the equation $E=MC^2$. This energy is transmitted radiatively to the star's surface, which is then radiated away.

If you've ever watched a fire burn, or seen metal being forged, you know that color is correlated with temperature. This is where the phrase "red hot" comes from. The peak wavelength of a star (a star will emit in various different wavelengths, just one wavelength is the most emitted) is governed by Wien's Law, which is $\lambda = b/t$, with b being the Wien's constant, which is 2.89777×10^{-3} meter-kelvin. This is why bigger stars are bluer, as their peak wavelengths are lower due to the increase in temperature (more gravitational pull creates higher fusion speeds). So, how does this relate to planetary detection? Stars have predictable peak wavelengths that can be predicted and measured. Because of the laws of gravitation, when one object is orbiting another, as in the case of most planetary systems, it is actually the case that they are orbiting the Barycenter of the system, which is where the combined center of mass is. This causes the star to "wobble" ever so slightly. In this wobble, the star may move at speeds of around 50 m/s, which is detectable in the spectrum of the star, with the peak wavelengths shifting up and down in a cycle. The equation for radial velocity is given by.

$$V = \frac{c\Delta\lambda}{\lambda}$$

Where V is the radial velocity, c is the speed of light, $\Delta\lambda$ is the change in wavelength, and λ is the "rest" wavelength. Through the radial velocity determined by the wavelength of the star, mass estimates of the star, and the time period of the star, the mass and orbital distances of the exoplanet can be determined through the relationships

$$M_1 V_1 = M_2 V_2$$

Where $M_1 V_1$ is the mass and velocity of one object, and $M_2 V_2$ is the mass and velocity of the other orbiting body and

$$p^2 = \frac{4\pi^2}{G(m_1 + m_2)} a^3$$

Where p is the orbital period, and a is the Semi-Major axis of the orbiting body. The m_1 and m_2 are the respective masses of the parent and orbiting body, and G is the Universal Gravitational Constant 6.67×10^{-11}.

In the experiment, we will be using sound waves instead of light, as spectrum equipment is very expensive. This is a demonstration of the doppler effect.

4.13   Torch Spectroscopy by Brian Lu

Materials:

1. Propane Bunsen burner
2. Various metal salts (e.g. sodium chloride, potassium chloride)
3. Water
4. Popsicle sticks

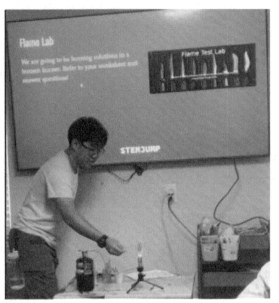

Procedures:

1. Dissolve the salts at maximum dissolvable concentration in some water, and soak the popsicle sticks inside (leave half of them dry).
2. Turn on the burner.
3. Burn the soaked part of the sticks with the burner, watch the color!
4. Turn off the burner.
5. Soak the burnt sticks in water to keep fire risk down.
6. Let the sticks cool down.

Safety:

- Always keep a trained adult in the room when performing this experiment; **Fire is very hazardous!**
- The salts can sometimes produce significant amounts of heat when being dissolved. When dissolving salts, add water to the correct amount and slowly pour in salt until dissolved. If done incorrectly, it may get very hot.
- Salt solutions may splash - may cause irritation
- You may want to have gloves, as some of these salts will irritate the skin.

Experiment Questions:

1. Why do these solutions glow different colors?
2. What application might this have to us in day-to-day life?
3. What application does this have to Astronomy?

Background and Applications:

Have you ever wondered how we know the compositions of things, stars, planets, that are millions of miles away? How do we know that the exoplanet K2-18b has byproducts and building blocks of life in its atmosphere? Spectroscopy, of course! Spectroscopy is essential to Astronomy as a field, but has many applications throughout the rest of STEM.

So, how does Spectroscopy work? Before we get into that, we need to talk about Kirchoff's Gas Laws. Kirchoff's Gas Laws describe how gasses behave when electromagnetic radiation interacts with them. Kirchoff's laws describe 3 types of gasses: hot, dense gasses; hot, un-dense gasses (is there even a word for that?); cold, less dense gasses, as shown below.

Hot, dense gasses emit a continuous spectrum of light with no gaps. This is because they approximate blackbodies, meaning that they absorb all frequencies of light perfectly, at any angle, and at any wavelength. No body can practically emit all wavelengths at an equal power, with most stars having a wavelength peak dependent on their effective temperature.

While blackbodies by themselves are not particularly useful in helping us determine the composition of certain things, it is helpful when the light emitted passes through a colder, less dense gas cloud that is not emitting anything? such as an atmosphere (all bodies emit radiation unless they're at absolute zero, but we can simplify them down as not emitting because their emissions are long wavelengths, low powers, and are easily filtered). After the gas cloud has passed through, we see that the spectrum has had lines cut in them, with certain wavelengths being absorbed by the gas cloud that we've passed light through.

This is how we determine the composition of exoplanet atmospheres, where, by waiting for transits (meaning that the planet passes in front of its parent star), and looking for light from the star that has passed through the exoplanet's atmosphere, we can determine its composition by comparing its spectrum to the known spectrums of elements we have tested here on earth.

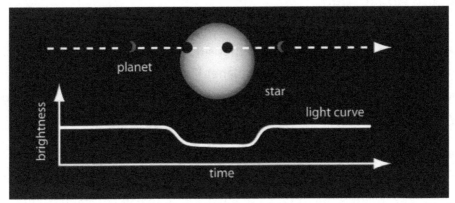

Figure 4.13-1 A combined diagram of the graph of a star's light curve during a planetary transit, and a visual depiction of a planetary transit. Source: NASA

Finally, we come to the hot and un-dense gas cloud. Like the hot dense gas, it has enough thermal energy to emit light that we can see; however, it is not dense enough to approximate a blackbody, so instead of emitting a constant spectrum, it emits in specific wavelengths.

The fundamental reason why all of these occur is concepts dealing with electron orbitals, energy, and quantum physics/mechanics, but at our current level, all we need to remember is that different gasses absorb/emit unique wavelengths of light.

Now, let's talk about the experiment that we are going to perform. We (and probably neither do you) don't have the experience nor funding to purchase and use a $35000 spectroscopy machine, so we will improvise and experiment using Kirchoff's hot but un-dense gas. Each of the metal ions present in the salts that were listed earlier have unique emission spectrums when vaporized at high enough temperatures. So, when you burn the sticks soaked with the solution, it will have enough energy to vaporize the solution and ionize it (that is what the fire is, plasma), creating the scenario that we discussed before.

Chapter 5 Engineering

5.1 Catapult Making by Brian Lu

Materials:

1. Popsicle sticks
2. Tape
3. Rubber bands

Procedures:

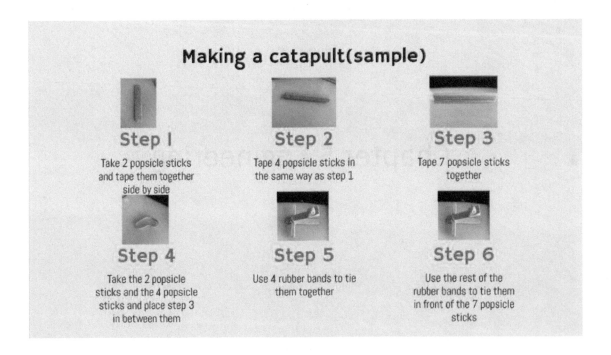

Making a catapult(sample)

Step 1
Take 2 popsicle sticks and tape them together side by side

Step 2
Tape 4 popsicle sticks in the same way as step 1

Step 3
Tape 7 popsicle sticks together

Step 4
Take the 2 popsicle sticks and the 4 popsicle sticks and place step 3 in between them

Step 5
Use 4 rubber bands to tie them together

Step 6
Use the rest of the rubber bands to tie them in front of the 7 popsicle sticks

Safety:

Don't poke your eyes with popsicle sticks.
Don't shoot yourself with the catapult.

Experiment Questions:

1. Why will a catapult never exceed 60% efficiency?
2. What is mechanical advantage?
3. What is potential energy? Kinetic energy?

Background and Applications:

Humans have launched rocks since the dawn of time; first with our arms, then with crude sticks, then with catapults, to today where we use chemical explosives to launch metal rocks through the sky. In medieval times, the goal of catapults was to launch large rocks as far as possible in order to destroy fortifications. But how do we ensure that the rocks go as far as possible?

Converting potential energy to kinetic energy as efficiently as possible is the goal of our catapult. In the catapults you will be building, energy is stored in the form of elastics. Most catapults in real life store energy in the form of gravitational potential energy, dropping counterweights to fling rocks at extremely high speeds.

Keeping the weight side of the arm short and lengthening the projectile side of the arm creates the maximum mechanical energy being transferred.

We can calculate the maximum theoretical distance a projectile can be launched by calculating the potential energy of the weight with the equation PE = mgh, m = mass, g = gravitational acceleration, and h = height. We then assume that half of this energy is used to propel the mass vertically. Using this, we can calculate how long the project stays in the air using F = MA. Then, using the time the projectile stays in the air, we can calculate the distance the projectile travels. This is the theoretical maximum distance it can travel. In practice, a catapult will never exceed about 60% efficiency.

5.2 Paper Airplane by Claire Long

Materials:

1. Paper

Procedures:

Note: There are many ways to fold a paper airplane, but we will be working with the most basic one.

1. Fold the paper in half (Hotdog Style), so that the long edges touch each other.
2. Unfold the paper so that there is a crease running down the center of the paper. Fold one corner down to the center line, and do the same to the other corner.
3. Rotate the paper so that the folded part is on top.
4. Fold the new top edges (the two folds you just created) into the center line as well. You should have a triangular shape.
5. Fold the plane in half along the centerline so that the folds are inside the paper.
6. Fold the wings down so that they meet the bottom edge of the plane's body.
7. Unfold it so that the wings are perpendicular to the body, which you will use to throw.
8. Throw it as hard as you can!!

Safety:

Do not throw the airplane at anyone, please!

Experiment Questions:

1. What is lift?
2. What is drag?
3. How does the paper airplane fly?
4. What is an airfoil and how does an airfoil work?

Background and Applications:

Paper airplanes use aerodynamics to fly. There are four main forces of aerodynamics: **Lift, Drag, Thrust, and Weight**. **Lift** is the force that causes any object to move upwards. Air moves over and under the wings, which pushes the plane upwards and causes it to fly. **Thrust** is a type of force that pushes the plane along. When you throw the paper airplane, it pushes it forward, which is an example of thrust. **Drag** is the force of the air that pushes against the paper airplane when it flies and is the opposite of lift. When the air simply pushes against the airplane and does not flow around it, this is called drag and slows the plane down. Finally, the **weight** of the plane, or how much mass it has, causes the plane to steadily move downwards with gravity and eventually makes it fall down.

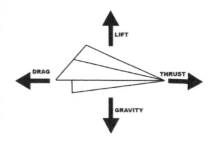

Figure 5.2-1 The four forces of aerodynamics on an airplane and their directions

Lift in paper airplanes is caused by the interaction between the wings and the air around them. While not perfect, paper airplanes simulate real airplanes. Real airplanes have **airfoils, or** teardrop shaped wings, which generate lift. Due to the longer distance the air has to travel on the top of the airfoil, which has a curved shape, the air travels faster with a lower pressure. Meanwhile, on the bottom of the airfoil, the air travels slower. According to the **Bernoulli Principle,** as the speed of a fluid increases, the pressure will decrease. Because of this, as air travels faster, the pressure decreases, and vice versa. Due to this rule, the slower air on the bottom of the airfoil has a higher pressure. Since molecules want to move from areas of high pressure to low pressure, this creates a force upwards, known as the lift force. The tapered end helps to seamlessly connect the two regions of air pressure - if the wing had just ended in a flat surface, the two pressures would mix, and cause turbulence and drag. With the tapered end, the air flows evenly and smoothly.

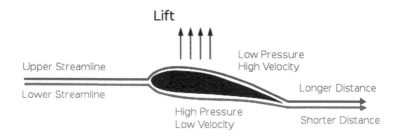

Figure 5.2-2 Forces and flow acting on an airfoil. Source: https://en.wikipedia.org/wiki/Lift_%28force%29

5.3 Solar System Model by Brian Lu

Materials:
1. Paper of cutouts of each planet *Measurements @ the end
2. 1 set of colored pencils
3. 1 pencil
4. 1 pair of scissors

Procedures:
1. Cut out each of your planets.
2. As we move along, you will have ten minutes to color your planet and write down facts about it on the back of your planet.
3. In this order: Sun, Mercury, Venus, Earth, Mars, Asteroid belt, Jupiter, Saturn, Uranus, Neptune, Kuiper Belt.

Safety:
Do not poke yourself when using scissors.
Do not run with scissors in your hand.

Experiment Questions:
1. Why is accurately modeling our solar system difficult?
2. What is an "AU"?

Background & Application
When you look up into the night sky, we see hundreds of thousands of stars. Thousands of years ago, the ancient Greeks looked up and saw what we see today (with a few differences due to earth's precession, supernovas, and whatnot), and they named these points of light "Ἀστήρ." These points of light move in predictable patterns throughout the year, and the ancient Greeks picked up on this. However, they also noticed six other points of light that did not adhere to these rules. The Greeks named these "λανήτης (planētēs)," which meant "wanderer." The Greeks did not know the mechanisms behind the "wanderers," and humanity would not know until nearly two thousand years later, when Copernicus would publish his Heliocentric theory in 1543.

The Solar System taught today in schools consists of 8 planets, the sun, various moons, and asteroid belts. Starting closer to the sun are the terrestrial planets Mercury, Venus, Earth, and Mars. Then, we see the asteroid belt, which separates the rocky planets from the Gas giants, Jupiter, Saturn, Uranus, and Neptune, which inhabit the outer "half" of the Solar System. We say "half" because of the way that the Solar System is depicted in media and in most educational diagrams, which shows Jupiter being about halfway between Neptune and the sun, which is incorrect.

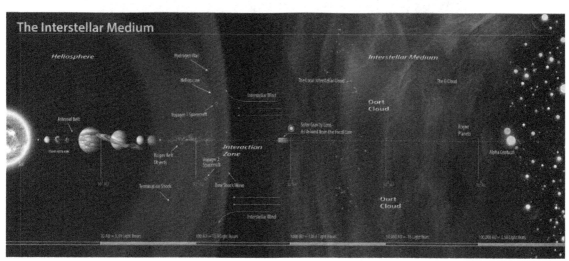

Figure 5.3-1 A artistic depiction of the planets of the Solar System. Source: By Charles Carter/Keck Institute for Space Studies - https://web.archive.org/web/20190617075031/https://www.nasa.gov/feature/jpl/interstellar-crossing-the-cosmic-void/. (See also https://kiss.caltech.edu/final_reports/ISM_final_report.pdf), Public Domain, https://commons.wikimedia.org/w/index.php?curid=147469703

While this diagram is not as egregious as some others, which depict the planets not to scale, this diagram does not show the immense distances between the planets. The earth is 1 AU from the sun (on average), or 1.496×10^{11} meters. This distance is so vast that it takes light, and gravity, nearly nine minutes to traverse it. Jupiter, however, is 5.2 AU from the sun, and Neptune is 30 AU from the sun.

Now, you may ask, why is the Solar System arranged this way, why are all the Rocky planets closer to the sun and all the Gas giants farther away? This is due to how the sun formed. Before the sun was born, it was just a cloud of pretty loose hydrogen gas. Some event (like a nearby Supernova) triggered some of the gas to start collapsing into a new star. As the star progresses towards nuclear fusion, surrounding gas and dust are blown away by the Solar wind of the star, with the lighter gasses of the Gas giants too light and too hot to condense into planets, while some more dense rocky elements stay behind, creating the rocky planets.

Figure 5.3-2 An artistic depiction of the formation of a star. Source: NASA

With this experiment, we will be building a Model of our Solar System. Whether it is to scale or not is up to you. Below, you can find some measurements for the planet's radius that can be used. Keeping the distances between planets to scale is very difficult and would make smaller planets like mercury literally microscopic if you tried to fit it into a poster board; however, you can model accurate distances with the cutouts before attaching them to the board.

	Sun	Mercury	Venus	Earth	Mars	Jupiter	Saturn	Uranus	Neptune
	Radius (inch)								
22" x 28"	14	0.05	0.12	0.13	0.07	1.44	1.17	0.51	0.495

5.4 Water Filter by Claire Long

Materials:

1. Filter paper (coffee filter)
2. Sand - 1 cup
3. Perler beads - 1 cup
4. Orbeez - 1 cup
5. Plastic bottle

Procedures:

1. Cut off the bottom of the plastic bottle and flip the bottle upside down.
2. Add filter paper, perler beads, orbeez, and sand, in whatever order, into the plastic bottle.
3. Get something to support the top of the bottle and catch the water. We like to put the inverted top of the plastic bottle back into the body, but using a cup works as well.
4. Add water!
5. Continue to experiment with the order of materials, and maybe try adding a few materials of your own!

Safety:

Don't drink the water! It is technically safe, but it will not be clean enough.

Experiment Questions:

1. What order of materials will filter the water most efficiently?
2. Why is it important to layer the materials in the correct order?
3. How does this filter work?
4. Can you think of any other common materials that could be used for a filter?
5. Is the filter capable of eliminating all foreign particles?

Background and Applications:

All these materials we used in the experiment are able to filter water through different means. Water often contains many large particles called impurities: dirt, sand, and even some bacteria are examples. Water is able to flow through small gaps, called pores, relatively easily- however, larger particles and pathogens sometimes cannot. Filter paper, sand, orbeez, and perler beads all work on this principle: as the water flows through the pores, the larger particles get trapped between the substance. The effectiveness can vary depending on the size of the pores, with sand having the finest and smallest holes of them all, trapping the particles between sand grains. Orbeez also have a separate way of trapping particles, having the ability to expand (check out the page on orbeez in the physics section!), potentially swelling and trapping pollutants with the water it absorbs. The most effective layering system would be to decrease the size of the pores from where the water enters to where it exits: in this logic, you would use orbeez -> perler beads -> sand -> filter paper. The largest particles get caught at the top of the layer, and as it moves downwards, the finer particles get trapped in even smaller pores as it continues to flow through the filter.

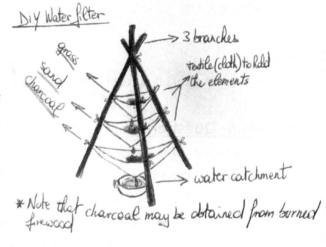

In the wild, other materials may be used, such as charcoal, sand, gravel, and small rocks. These can be layered to help create drinkable water if needed when out in the wilderness: alternate layers of materials with small and large pores, and then pour the water through. This makes the water clearer and more palatable, but boiling the water to sterilize it is still recommended to make it safer.

Figure 5.4-1 Example of a water filter. Source: https://commons.wikimedia.org/wiki/File:Homemade_waterfilter.jpg

5.5 Bridges by Emma Zeng

Materials:

1. Popsicle sticks (a lot)
2. Hot glue
3. Textbooks/books to test how much weight the bridges can hold

Procedures:

1. Using hot glue, glue the popsicle sticks together in a way that forms a bridge.
2. Place the completed bride across two tables with gap under the center of the bridge.
3. Place books one by one on the top of the bridge to see how much weight it can hold!

Safety:

When using the hot glue gun, have adult supervision. Don't touch the tip of the gun and be careful handling the hot glue.

Experiment Questions

- What would happen if we used different shapes? Triangles? Arches?
- What if we used a different material instead of popsicle sticks? Pipe Cleaners? Plastic Forks?
- Search up Truss Bridge. What makes it so stable?

Background and Applications:

Constructing bridges can be used to test theories, ideas and engineering skills. This activity is done through various grades, and for even adults! Building and testing these bridges can assess the functionality of real-life structures. Through this activity, we see the physics of bridges at play. The tension and compression forces of the popsicle sticks are what's strong enough to keep the structure up. Imagine a spring, if the two ends are pulled away from each other, there is a tension force created. On the other hand, if both ends of the spring are pushed together, there is a compression force created. The bridge is held up with these forces and needs to be built to sustain them. For example, in a suspension bridge, the material it's built with is pulled and stretched creating tension to hold weight. In an arch bridge, each arch has compression that pushes up, holding weight. Through designing and testing various ideas, this experiment can find different methods to provide stability to a bridge.

In real life, bridges and structures are also exposed to environmental conditions such as wind and rain. How would we build a bridge to sustain these conditions? We need to create a bridge that cannot bend or vibrate when struck with harsh winds, and one that also will not deteriorate under heavy rain. Engineers are tasked with this when designing and building bridges. Bridges actually need to be flexible. By this, I don't mean you could jump on it like a trampoline; instead, you would have a few inches of room to move. Many tall buildings are built similarly; they have a few inches to sway from side to side. For example, the world's current tallest building, the Burj Khalifa in Dubai, can experience up to a two-meter sway from its top floor in harsh winds! This way, when the wind hits the structures, rather than snapping them in half, the aerodynamic buildings and bridges bend slightly with the wind, allowing them to absorb the forces hitting it.

For Higher Grades:

The fundamental frequency is the lowest frequency of a periodic waveform. To put this into perspective, if we take a poorly designed bridge and have the wind blow it at its fundamental frequency, the entire bridge will bend and twist chaotically. Engineers are tasked to avoid these situations in order to construct a safe and functioning bridge. To start, they have to ensure the fundamental frequency of a bridge isn't in the range where it could be triggered by cars, bikes, or any natural environmental factors. To do so, they look at the equation $F1 = \frac{1}{2} V/L$. F1 is the fundamental frequency, V is the standing wave speed, and L is the length of the bridge or medium. So, for a bridge to avoid the range of 1.5HZ to 4.5Hz, the vibrations that most cars have, an engineer can change the length of the bridge to change its fundamental frequency. This way, bridges constructed can be strong and reliable through various conditions.

5.6 Circular Paper Plane by Emma Zeng

Materials:

1. Straws
2. Tapes
3. Paper
4. Scissor

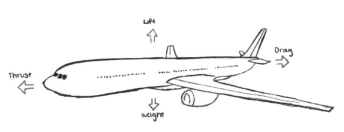

Figure 5.6-1 Airplane.

Procedures:

1. *You have 3 options when it comes to cutting the cardstock paper:*
   a. *Cut the cardstock paper so you have 3 strips, each piece measuring 5"x1". Take 2 of the strips and tape them together so now you have a 10"x1" strip.*
   b. *Cut the cardstock paper so you have 1 strip that is 10"x 1" and 1 strip that is 5" x 1".*
   c. *Print the **Straw Airplane Template** on cardstock paper and cut along the outlines for the 2 strips.*
2. *Curl the 5"x1" strip into a hoop and tape the ends together.*
3. *Repeat with the 10"x1" strip.*
4. *Tape the hoops to the two ends of the straw. Ensure the straw is lined up and on the inside of the hoops.*
5. *Hold the hoop glider in the middle of the straw with the little hoop in the front and throw!*

Safety:

Be careful when using scissors and other sharp objects.
When throwing the plane, be careful not to hit anyone.

Experiment Questions

1. Does the placement of the paper hoops on the straw affect the distance it can go?
2. Does the length of the straw affect the distance it can go?

Background and Applications:

This experiment examines aerodynamics. When building planes, it's essential to understand the four components that allow planes to fly. There is thrust, lift, drag, and weight. We can see these at play in our circle plane, enabling it to fly. The thrust is an arm or hand throwing the plane and force, allowing it to be pushed forward. In opposition to this, drag, commonly called air resistance, slows the plane down as it moves forward, counteracting the thrust. The lift is the air keeping the plane up, moving through the straw and hoops, allowing it to stay in the air. This is similar to the wings of airplanes, keeping it in the sky. In resistance to this, the last force weight is gravity. Gravity, as we all know, pushes objects back down to earth, causing our paper plane to land at the end of its flight.

The Wright Brothers created the first airplane in 1903. They used wings and propellers to create enough "thrust" and "lift." Today, airplanes are still modeled similarly.

We can see the contrasting forces at play, keeping the plane in the air. In the plane, thrust is the engine, drag is air resistance, lift is the air moving under the wings, and weight is gravity.

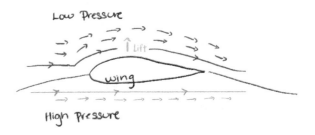

Figure 5.6-2 Airplane wing and the air moving by it.

Let's focus on the wings of the plane. This is an essential component that keeps the plane in the sky. Airplane wings are shaped in a specific way so that the air on top will move faster. Mention the actual shape and why the air moves faster on one side as compared to the other? The faster air moves, the lower the pressure. Because the pressure of the air moving under the wing is now higher, it creates a force that pushes the wing up, keeping the plane in the sky.

5.7 Bottle Rocket by Emma Zeng

Materials:

1. Empty plastic bottle of some kind, ex: 2L soda bottle
2. Cardboard and construction paper for fins
3. Hot glue/tape
4. Something to cut the bottle and cardboard (Knife/Scissor)
5. Bottle rocket pump

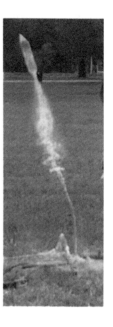

Procedures:

1. Cut out 3-4 fins in the shape of triangles.
2. Attach to the base of the bottle using tape, in this case it's the cap end.
3. Using a sheet of construction paper, wrap around the base of the bottle/the opposite side of fins, into a cone shape. This will act as the tip of the rocket.
4. Fill the bottle a quarter of the way with water.
5. Using the pump, pump the bottle with air.
6. Launch! When suddenly releasing the air, the bottle will shoot up.

Safety:

When using a sharp knife or scissors, have adult supervision.

Do not do this activity indoors.

When launching the bottle, avoid hovering over it to look at it.

Once the bottle is in the air, keep an eye on it, watch where it comes down, and avoid getting hit.

Experiment Questions

1. What would happen if we used bottles of different sizes?
2. What would change about the launch with more or less water?

Background and Applications:

In 1687, famous scientist and physicist Sir Isaac Newton published his three laws of motion. These laws solidified gravity and the predictability of an object's movements. Through this experiment, we learned about Newton's 3rd law of motion: Every action has an equal and opposite reaction. We can see this as air is pushed into the bottle by a pump, the pressure increases within the bottle. When we launch the bottle, and the pressure is suddenly released, the air is pushed with a strong force out of the bottle. Because of Newton's third law, as the bottle pushes the air out, releasing the build-up pressure, an equal and opposite reaction needs to occur. Therefore, the air pushes the bottle up with the same amount of force in the opposite direction. In simple terms, as the bottle pushes air down, air pushes the bottle back. Through this, the bottle is launched into the air as the rocket we see.

So, what would happen if we added more air to the bottle by increasing the pressure? Essentially, increasing pressure increases the resulting force. This goes into Newton's second law: Force (F) is equal to mass (kg) times acceleration (m/s), also written as $F = m \times a$. If we increase the force and keep the same mass, we would end up with a larger acceleration. Moving elements of Newton's second law around, we get a = f/m, acceleration is equal to force divided by mass. In other words, a larger force and a smaller mass would result in a larger acceleration. We can experiment with this equation by changing the amount of water we put in the bottle, air we pump into it, and size of the bottle.

Years before Sir Isaac Newton published his findings, two other scientists, Johannes Kepler and Galileo Galilei had been deep in researching planetary movement and gravity. Newton linked these findings and demonstrated terrestrial physics' relation to planetary movement. He was able to prove that the force that acted on planets and moons from previous studies was the same force that let an apple fall from a tree. (The fact is, the tale of an apple falling on Newton's head is not true). So now that we have established the law of gravity, and have seen it in action, how do we get rockets into space?

To put it simply, a rocket must have enough thrust to counteract earth's gravity. With enough propellants that cause a large enough thrust, the rocket is pushed upwards with a force greater than gravity's force pulling it down.

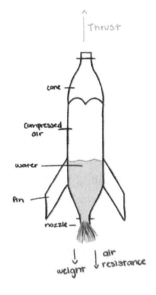

Figure 5.7-1 Bottle rocket

Acknowledgments

We extend our heartfelt thanks to everyone who contributed to the creation of this book.

Thank you to our teachers and experiment creators from STEMJUMP's Summer Camp 2023: Ray Chen, Evan Huang, Audrey Li, Claire Long, Karen Wang, Michael Xia, Hannah Xie, Alan Yue, and Alex Zhang. Your dedication to lesson planning, creating instructions, and overseeing the initial experiments fostered an amazing camp experience. Your discoveries are the foundation for this comprehensive guide of scientific activities.

We are also deeply grateful to our current board members: Carter Feng, Claire Long, Christopher Ou, Karen Wang, Hannah Xie, Emma Zeng, and Eric Zhang. Your contributions, from transcribing the experiments to elaborating on each activity, were invaluable. Your confidence in this book allowed us to push past any obstacles and reach publication.

Additionally, we want to express our gratitude to STEMJUMP's Day 1 supporters: all of our parents. From seemingly trivial tasks like driving us to meetings, to your unwavering presence as our backbone and greatest advisors, you have enabled us to achieve STEMJUMP's mission.

Most importantly, we express our sincerest thanks to all of our students and their parents. You make STEMJUMP possible. The smiles on your faces after learning a new concept and the excitement in your voices when discussing experiment results fuel our passion for STEMJUMP's success. We hope this book brought these smiles and excited voices into your home.

Thank you all for making this book possible. We cannot wait for STEMJUMP's next endeavor, motivated by your continued support.

STEMJUMP Co-Founders, Brian Lu and Jessica Fan
June, 2024

Made in the USA
Coppell, TX
21 July 2024

35028491R00103